Nelson Yuan-sheng Kiang

HANDBOOK OF PROBIOTICS

HANDBOOK OF PROBIOTICS

YUAN-KUN LEE
National University of Singapore
Singapore

KOJI NOMOTO
Yakult Central Institute for Microbiological Research
Tokyo, Japan

SEPPO SALMINEN
University of Turku
Turku, Finland

SHERWOOD L. GORBACH
Tufts University
Boston, Massachusetts

A Wiley-Interscience Publication
JOHN WILEY & SONS, INC.
New York • Chichester • Weinheim • Brisbane • Singapore • Toronto

This text is printed on acid-free paper. ♾

Published simultaneously in Canada.

Library of Congress Cataloging in Publication Data:

Handbook of probiotics / Yuan-Kun Lee . . . [et al.].
p. cm.
Includes index.
ISBN 0-471-19025-X (cloth : alk. paper)
1. Intestines—Microbiology—Handbooks, manuals, etc. 2. Food—Microbiology—Handbooks, manuals, etc. 3. Microorganisms—Therapeutic use—Handbooks, manuals, etc. I. Lee, Y. K. (Yuan Kun)
[DNLM: 1. Bacterial Physiology. 2. Probiotics—pharmacology. 3. Probiotics—therapeutic use. 4. Food, Formulated. QW 52H236 1999]
QR171.I6H36 1999
664'.001'579—dc21
DNLM/DLC
for Library of Congress 98-27511

Printed in the United States of America

10 9 8 7 6 5 4 3 2 1

CONTENTS

PREFACE

Emerging clinical evidence of health benefits and a better understanding of the scientific basis of probiotics have caught the interest and imagination of research scientists and end users alike. We are witnessing a period of rapid expansion in the market for probiotic products, particularly in Europe and Asia. Many of the traditional "health foods" in the form of indigenous fermented foods and beverages have been relaunched and the concept of probiotics is being applied to farm animals and aquaculture.

Books on probiotics published so far have been mainly concerned with the tasks of explaining what probiotics are and how they work. Probiotics have come of age. There is a need for a "data book" as a source of technical information on probiotics. In this book, representative research and clinical works were selected and the data presented in user-friendly forms for rapid reference and easy extraction of information.

In this Handbook, "probiotics" is defined as viable bacterial cell preparation or foods containing viable bacterial cultures or components of bacterial cells that have beneficial effects on the health of the host. Thus, live microorganisms and substrates (sometimes termed prebiotics) that promote growth of indigenous probiotic microorganisms, as well as components of probiotic microorganisms, are included.

It is the aim of this Handbook to put together data and methodology required in the development of a successful probiotic product from laboratory to the marketplace and to provide information on the choice and application of the appropriate probiotic for a particular purpose. This book will serve as a resource material for students and researchers.

THE AUTHORS

HANDBOOK OF PROBIOTICS

1

INTRODUCTION

1.1 PROBIOTICS

Probiotics are viable bacterial cell preparation or foods containing viable bacterial cultures or components of bacterial cells that have beneficial effects on the health of the host (1, 3, 5, 6, 8). Many of these probiotics are lactic acid bacteria (see Section 1.2). Probiotic lactic acid bacteria are useful in the treatment of disturbed intestinal microflora and increased gut permeability, which are characteristic to many intestinal disorders (4). Examples include acute rotavirus diarrhea, other intestinal dysfunction, subjects with food allergy, subjects with colonic disorders, and patients undergoing pelvic radiotherapy (see Chapter 4). In such disease states altered intestinal microflora, impaired gut mucosal barrier, and intestinal inflammation may be present. Other properties of specific probiotic lactic acid bacteria include modification of intestinal microflora and its metabolic products, such as short-chain fatty acids and antimicrobial components (see Chapter 5). Successful probiotic bacteria are able to survive gastric conditions and colonize the intestine, at least temporarily, by adhering to the intestinal epithelium (see Chapter 2). Well-documented probiotic strains are clearly characterized and clinically documented. Such probiotic microorganisms with demonstrated probiotic properties appear to be promising candidates for the treatment of clinical conditions with abnormal gut microflora and altered gut mucosal barrier functions. They are also promising components for future functional foods and clinical foods for specific disease states provided that the basic requirements for new strain selection, safety assessment, and clinical studies are carefully followed. Currently, the European Union is supporting research to demonstrate the health benefits of probiotic lactic acid bacteria not only to the common consumer but also for infants, children, and the elderly. For this purpose, the stabilization of normal intestinal function and the treatment of specific dysfunction involving intestinal microflora and gut mucosal barrier are assessed.

Similar effort is fostered in Asia, in particular Japan, supported mainly by private organizations.

The human gastrointestinal tract exhibits variability in bacterial numbers and composition between the stomach, small intestine and colon (2). The total bacterial count in gastric contents is usually below 10^3 per gram, with reduced numbers due to the low luminal pH. In the small intestine, bacterial numbers range from approximately 10^4 per milliliter of contents to about 10^6–10^7 at the ileocecal region (2). The main factor limiting growth in the small bowel is the rapid transit of contents, as well as certain secretions such as bile. The human large intestine is an intensely populated microbial ecosystem (3), with typical numbers of about 10^{11}–10^{12} for every gram of contents. Several hundred different species of culturable bacteria are thought to be present in the large intestine under normal circumstances, the vast majority of which are strict anaerobes. The composition is able to respond to anatomical and physicochemical variations that are present. The right (proximal) colon is characterized by a high substrate availability and low pH. The left, or distal, area of the colon has a lower concentration of available substrate, the pH is approximately neutral, and bacteria grow more slowly.

During the recent decade a major change has taken place in the way in which the biological activities of the human colon are viewed (4). It has been observed that the human colon is an intense area of metabolic activity that has an important role in digestion, with many of the functions attributed to the resident flora. The functions and dysfunctions of intestinal microflora offer the place for correcting or attempting to correct the balance of human intestinal microflora with systemic beneficial health effects for the host. For this purpose, probiotic bacteria have been applied per os (4, 9).

There are currently still relatively few well-designed and well-conducted human intervention studies with probiotics. These indicate significant differences between probiotic bacteria and strains. A recent review by a European working group on probiotics and prebiotics listed several effects as scientifically proven at least for some probiotic strains (Table 1.1).

As it is clear that all reported effects are strain dependent, even slight differences in the strain properties influence the nutritional and clinical outcome. Thus, it is important that each strain is tested on its own or in products designed for the particular

Table 1.1 Scientifically Established Health-Related Effects of Specific Probiotic Strains (2, 7, 8, 9, and Chapter 4)

Alleviation of symptoms of lactose intolerance
Immune enhancement
Shortening of the duration of rotavirus diarrhea
Decreasing fecal mutagenicity
Decreasing fecal bacterial enzyme activity
Prevention of recurrence of superficial bladder cancer

Table 1.2 Requirements for Clinical Studies of Lactic Acid Bacteria to Verify Effects on Gut Mucosal Barrier (1, 4, 6, 8, 9).

Each strain documented and tested independently
Extrapolation of data from closely related strains not acceptable
Well-defined probiotic strains, well-defined study products, well-defined study populations
Double-blind, placebo-controlled and randomized human studies
Results confirmed by several independent research groups
Publication in peer-reviewed journals

function. For this purpose, the guidelines for clinical studies should be carefully followed (Table 1.2). However, it is also important to provide quality assurance for the products and strains to ensure that the properties of each strain are stable and do not alter during long periods of industrial utilization.

In conclusion, lactic acid bacteria appear to have some documented beneficial effects on gastrointestinal disturbances including some forms of diarrhea and related diseases associated with intestinal dysbiosis. Many of these diseases share common features, which are related to altered gut mucosal barrier functions. Probiotic bacteria offer new dietary alternatives for the management of these conditions through stabilization of the intestinal microflora, promotion of colonization resistance, and preservation of intestinal integrity. In addition, lactic acid bacteria may be important in preserving the intestinal integrity and promoting colonization resistance. The immunomodulatory properties of probiotics indicate that they may also have a role in the dietary management of food allergy.

The aim of this book is to provide a background of probiotics for research purpose and for the selection of novel probiotics, as well as for the design of nutritional and clinical studies for verifying the effects of probiotic microorganisms in humans, both normal consumers and populations with different intestinal dysfunctions. Our aim is to assist in the development and further progress of this important field of functional ingredients and functional foods.

REFERENCES

1. Fuller, R. Probiotics in man and animals. *J. Appl. Bacteriol.* **66:** 365–378, 1987.
2. Gorbach, S. L., Nahas, L., and Lerner, P. I. Studies of intestinal microflora. I: Effects of diet, age and periodic sampling on numbers of faecal microorganisms in man. *Gastroenterology* **53:** 845–855, 1967.
3. Havenaar, R., and Huis In't Veld, J. H. J. Probiotics: A general view. In *The Lactic Acid Bacteria* (B. J. B. Wood, ed.), Vol. 1, pp. 151–170. Elsevier, London, 1992.
4. Lee, Y. K., and Salminen, S. The coming of age of probiotics. *Trends Food Sci. Technol.* **6:** 241–245, 1995.
5. Lilly, D. M., and Stillwell, R. H. Probiotics: Growth promoting factors produced by microorganisms. *Science* **147:** 747–748, 1965.

6. Parker, R. B. Probiotics, the other half of the antibiotic story. *Anim. Nutr. Health* **29:** 4–8, 1974.
7. Salminen, S., Deighton, M., Benno, Y., and Gorbach, S. L. Lactic acid bacteria in health and disease. In *Lactic Acid Bacteria; Microbiology and Functional Aspects* (S. Salminen, and A. von Wright, eds.), pp. 211–253, Marcel Dekker Inc., New York. 1998.
8. Salminen, S., Gibson, G., Bouley, M. C., Isolauri, E., Boutron-Rualt, M. C., Cummings, J., Franck, A., Rowland, I., and Roberfroid, M. Gastrointestinal physiology and function: The role of prebiotics and probiotics. *Br. J. Nutr.* **80,**Suppl. 1: 147–171, 1998.
9. Salminen, S., Isolauri, E., and Salminen, E. Clinical uses of probiotics for stabilizing the gut mucosal barrier: Successful stains and future challenges. *Antonie van Leeuwenhoek* **70:** 347–358, 1996.

1.2 PROBIOTIC MICROORGANISMS

1.2.1 Microorganisms Used as Probiotics

Taxonomic Nomenclature[a,b]	Former Designations[a,b]	Optimum Temperature (°C)	Fermentation Products	Supplier[c]
Bifidobacterium breve	*B. parvulorum*	37–41	Lactic acid, acetic acid, formic acid	Strain Yakult: Yakult, Tokyo, Japan
B. bifidum	*Bacillus bifidus communis, Bacillus bifidus, Bacteroides bifidus, Bacterium bifidum, Tissieria bifida, Nocardia bifida, Actinomyces bifidus, Actinobacterium bifidum, B. bifidum, Lactobacillus bifidus, L. parabifidus, Cohnistreptothrix bifidus, Actinomyces parabifidus*	37–41	Lactic acid, acetic acid, formic acid	Strain YIT 4002: Yakult, Tokyo, Japan. Strain A234-4: Japan Bifidus Foundation, Tokyo, Japan
B. infantis	*B. liberorum, B. laetentis, B. parabifidum*	37–41	Lactic acid, acetic acid, formic acid	Strain ATCC15697: ATCC, Rockville, MD, USA
B. lactis	*B. animalis*	37–41	Lactic acid, acetic acid, formic acid	Chr. Hansen Ltd., Hørsholm, Denmark
B. longum	*B. longum* subsp. *animalis*	37–41	Lactic acid, acetic acid, formic acid	Strain F6-1-ES, 69- 2bs: Nikken Chemicals, Tokyo, Japan
B. thermophilum	*B. ruminale*	37–41	Lactic acid, acetic acid, formic acid	Strain P2-91

(continued)

Taxonomic Nomenclature[a,b]	Former Designations[a,b]	Optimum Temperature (°C)	Fermentation Products	Supplier[c]
Lactobacillus acidophilus	*Bacillus acidophilus, Thermobacterium intestinale*	35–38	Lactic acid	Strain NCDO1748: National Collection of Dairy Organisms, Reading, UK Strain SBT2062: Snow Brand Milk Products, Kawagoe, Japan Strain NCFM: GPGurlock, Cincinnati, OH, USA
L. delbrueckii subsp. *bulgaricus*	*L. longus, Bacillus bulgaricus, Bacterium bulgaricum, Acidobacterium bulgaricum, Plocamobacterium bulgaricum, Bacterium biogurt, Thermobacterium bulgarium, Lactobacterium bulgaricum, L. bulgaricus*	40–43	Lactic acid	Strain FW148: Unilever Research Laboratory, Welwyn, UK Commercial Yogurt Starter Culture: Chr. Hansen's Laboratory, Milwaukee, WI, USA Marshall Products, Madison, WI, USA
L. helveticus	*Bacillus ε Bacillus casei ε Caseobacterium ε Thermobacterium helveticum, L. helveticum, Plocamobacterium helveticum, Lactobacterium helveticum*	45	Lactic acid	
L. casei	*Bacillus α, Bacillus casei α Caseobacterium vulgare, Bacterium casei α, Streptobacterium casei, Lactobacterium casei*	30–35	Lactic acid, acetic acid, ethanol, formic acid	Strain Shirota: Yakult, Tokyo
L. fermentum	*Bacillus δ , Bacillus casei δ, Lactobacterium fermentum, L. fermenti, Betabacterium jensenii, Bacterium gayoni, L. gayoni, Betabacterium longum, L. longus, Lactobacterium longum, L. cellubiosus*	45	Lactic acid, acetic acid, ethanol, formic acid	
L. johnsonii	*L. acidophilus*	35–38	Lactic acid	Strain LC1: Nestlé Ltd., Vevey, Switzerland

(*continued*)

Taxonomic Nomenclature[a,b]	Former Designations[a,b]	Optimum Temperature (°C)	Fermentation Products	Supplier[c]
L. plantarum	*Streptobacterium plantarum*	30–35	Lactic acid, acetic acid, ethanol, formic acid	Strain ATCC14917: ATCC, Rockville, MD, USA
L. rhamnosus	*L. casei* subsp. *rhamnosus*	37	Lactic acid, acetic acid, ethanol, formic acid	Strain GG: Valio, Helsinki, Finland Strain ATCC53103: ATCC, Rockville, MD, USA Lactophilus Laboratoires Lyocentre, France
L. salivarius	—	30–40	Lactic acid, acetic acid, ethanol, formic acid	Strain WB1004: Wakamoto Pharmaceutical, Tokyo, Japan
Enterococcus faecium	*Streptococcus faecium*	22–33	Lactic acid	Strain SF68: A B Cernell, Engelholm, Sweden; Giuliani, Lugano, Switzerland
Streptococcus salivarius subsp. *thermophilus*	*Streptococcus thermophilus*	37–45	Lactic acid	Strain ATCC 19258: ATCC Rockville, MD, USA Commercial Yoghurt Starter Culture: Chr. Hansen's Lab. Milwaukee, WI, USA; Marshall Products, Madison, WI, USA

[a]Butler, J. P. (ed.), *Bergey's Manual of Systematic Bacteriology,* Vol. 2. Williams & Wilkins, Baltimore, 1986.
[b]Balows, A., Truper, H., Devorkin, M., Harder, W., and Schleifer, K.-H. (eds.), *The Prokaryotes. A Handbook on the Biology of Bacteria: Ecophysiology, Isolation, Identification, Applications,* Vol. 2. Springer-Verlag, New York, 1992.
[c]Papers referred to in this handbook.

1.2.2 Detection and Identification of Lactic Acid Bacteria Using Molecular Biology Techniques

Detection Method	Target Gene	Species	Specificity	Ref.
Polymerase chain reaction (PCR)				
Species-specific primers based on respective species ribosomal ribonucleic acid (rRNA)	16S rRNA	*L. casei* *L. delbrueckii* *L. helveticus* *L. acidophilus*	Species specific	4
Species-specific primers designed from 16S rRNA	16S rRNA	*B. adolescentis* *B. longum* *L. acidophilus*	Species specific	27
Randomly amplified polymorphic deoxyribonucleic acid (RAPD) using a 10-mer primer	16S rRNA	*L. acidophilus* *L. helveticus* *L. casei* *L. reuteri* *L. plantarum* *E. faecium* *S. thermophilus*	Strain specific	2
RAPD using a 9-mer primer	None	*L. plantarum* *L. pentosus*	50% strain specific	10
Restriction fragment length polymorphism (RFLP)				
Gel electrophoresis of *Hin* dIII, *Eco* RI digests. A 420-base pair (bp) fragment of rRNA gene of *L. reuteri* DSM20016 used as probe	16S rRNA	*L. plantarum*	Species specific; recognize 8 strains	9
Pulsed-field gel electrophoresis (PFGE) of *Sma* I, *Apa* I digests	Total genome	*L. acidophilus*	Strain specific	22
Agarose gel electrophoresis of *Eco* RV, *Hin* dIII, *Bgl* II, *Eco* RV, *Bam* HI digests	Total genome	*L. delbrueckii* subsp. *delbrueckii* *L. delbrueckii* subsp. *lactis* *L. delbrueckii* subsp. *bulgaricus*	Species, subspecies, and strain	14
Agarose gel electrophoresis of *Asp* 718, *Cla* I,, *Eco* RI digests	Total genome	*L. reuteri* *L. plantarum*	Strain and variety specific	23
Southern hybridization				
Variable regions of 23S rRNA	23S rRNA	*L. acidophilus*	Strain	18

(*continued*)

Detection Method	Target Gene	Species	Specificity	Ref.
gene of *Escherichia coli* amplified by PCR with gene-specific primers, used as probes		*L. gasseri* *L. johnsonii*	specific	
Plasmid pNST43 DNA isolated from *E. coli* JM109 used as probe	rRNA	*L. acidophilus*	Strain specific	22
Three probes: 16S + 23S cDNA from *E. coli* MRE600, 16S rDNA from *E. coli* HB101, and 16S rDNA from *L. acidophilus* ATCC4356	16S rRNA	*L. acidophilus* *L. casei* *L. delbrueckii* subsp. *bulgaricus* *L. fermentum* *L. gasseri* *L. helveticus* *L. murinus* *L. plantarum* *L. salivarius* *L. reuteri* *L. delbrueckii*	Strain specific	20
A 2-kilo-base (kb) *Eco* RI DNA fragment from a plasmid in *L. helveticus* cloned into an *E. coli* vector, used as probe	*Eco* RI digested total DNA (RFLP)	*L. helveticus*	Species specific	19
A 50-kb cryptic plasmid from lactobacilli, designated profile type B used as probe in colony hybridization	Plasmid DNA	*L. fermentum* type B strains	Type strain specific	25
A 1.2-kb DNA probe isolated from genomic library of *L. curvatus* DSM20019 constructed in bacteriophage λ gt11	Unknown gene	*L. curvatus*	Species specific	16
Four probes derived from randomly cloned DNA fragments in plasmid vector pBR322 used in colony hybridization	Plasmid or chromosomal DNA	*L. acidophilus* O	Strain specific	21
Dot blot hybridization				
A cloned bank of *Eco* RI fragments of chromosomal DNA from *L. delbrueckii* subsp. *bulgaricus* type strain established in vector YRP17, used as probe	Leu-291 lesion gene	*L. delbrueckii* subsp. *delbrueckii* *L. delbrueckii* subsp. *lactis*	Species specific	3

(*continued*)

Detection Method	Target Gene	Species	Specificity	Ref.
		L. delbrueckii subsp. *bulgaricus*		
Reverse dot blot hybridization with captured probes	16S, 23S rRNA	Probe-specific lactic acid bacteria	Direct species-specific simultaneous identification for mixed population	6
Amplified *hdc* A genes from *L. buchneri* and *Leuconostoc oenos* used as probes	*hdc* A gene	Histidine decarboxylating lactic acid bacteria, e.g., *L. buchneri*	*Ldc* A gene specific	12
Comparison of DNA sequences				
Diagnostic regions revealed in a comparative analysis of aligned rRNA sequences used as target sites	16S rRNA	*L. brevis* *L. sanfrancisco* *L. reuteri* *L. farciminis* *L. pontis*	Species specific	26

Species-Specific Oligonucleotide Probes for the Identification of Probiotic Bacteria (1)

Probe	5′,3′ Sequence	Target	Specificity	Ref.
Yogurt starter bacteria				
pNST21		DNA	*S. thermophilus*	15
I41	Corresponds to bp 2362–2894 of *E. coli*	23S rRNA	*S. thermophilus*	13
	GGG ATG GTA AAG AAT TTC TT	DNA	*S. thermophilus*	13
	CGA TGG TAA AGA ATT CCT TC	DNA	*S. thermophilus*	13
	GAT CGT AAA GAA TTC CTT CG	DNA	*S. thermophilus*	13
	ATG GTA AAG AAT TCC TTC GT	DNA	*S. thermophilus*	13
	GCT AAA GAA TTC CTT CGT	DNA	*S. thermophilus*	13
	GGT AAA GAA TTC CTT CGT CC	DNA	*S. thermophilus*	13
st	GTA AAG AAT TCC TTC GTC CA	DNA	*S. thermophilus*	13
	TAA AGA ATT CCT TCG TCC AC	DNA	*S. thermophilus*	13
Sth	CAT GGC CTT CGC TTA CGT C	23S rRNA	*S. thermophilus*	7
p21	5.5 kb plasmid	DNA	*L. delbrueckii*	24
pY85	Complementary to leu-291 lesion of *E. coli* strain CE891	DNA	*L. delbrueckii*	3
sLH1	*Hin* dIII fragment	DNA	*L. helveticus*	17
sLH2	*Hin* dIII fragment	DNA	*L. helveticus*	17
sLH2	*Hin* dIII fragment	DNA	*L. helveticus*	17
	Eco RI fragment f of pCG36	DNA	*L. helveticus*	19

(*continued*)

Species-Specific Oligonucleotide Probes *(cont.)*

Probe	5′,3′ Sequence	Target	Specificity	Ref.
Lactobacilli				
Lba(1)	TCT TTC GAT GCA TCC ACA	23S rRNA	*L. acidophilus*	18
ba(2)	AGC GAG CUG AAC CAA CAG AUU C rRNA probe	16S rRNA	*L. acidophilus*	7, 8
Lbcp	CAA TCT CTT GGC TAG CAC	23S rRNA	*L. crispatus*	5
Lbg	TCC TTT GAT ATG CAT CCA	23S rRNA	*L. gasseri*	18
Lbj	ATA ATA TAT GCA TCC ACA G	23S rRNA	*L. johnsonii*	18
Lbcr	GCA GGC AAT ACA CTG ATG	23S rRNA	*L. casei/ rhamnosus*	8
Lbcrp	CTG ATG TGT ACT GGG TTC	23S rRNA	*L. casei/ paracasei/ rhamnosus*	8
Lbpa	CAC TGA CAA GCA ATA CAC	23S rRNA	*L. paracasei*	8
Lbr	GAT CCA TCG TCA ATC AGG	16S rRNA	*L. reuteri*	6
	Plasmid DNA probe		*L. reuteri*	24
Lbfe (1)	GCG ACC AAA ATC AAT CAG G	16S rRNA	*L. fermentum*	25
Lbfe(2)	AAC GCG UUG GCC CAA UUG AUU G	16S rRNA	*L. fermentum*	7
Lbp	AAC GAA CUA UGG UAU UGA UUG G	16S rRNA	*L. plantarum*	7
Lbpp	ATC TAG TCG TAA CAG TTG	23S rRNA	*L. plantarum/ pentosus*	8
Bifidobacteria				
Bif64	CAT CCG GCA TTA CCA CCC	16S rRNA	*Bifidobacterium* (83 ± 2%)	11
Bif662	CCA CCG TTA CAC CGG GAA	16S rRNA	*Bifidobacterium* (73 ± 4%)	11
Bif1278	CCG GTT TTC AGG ATC C	16S rRNA	*Bifidobacterium* (34 ± 4%)	11
PAD	GCT CCC AGT CAA AAG CG	16S rRNA	*B. adolescentis*	28
PBI	GCA GGC TCC GAT CCG A	16S rRNA	*B. bifidum*	28
PBR	AAG GTA CAC TCA ACA CA	16S rRNA	*B. breve*	28
PIN	TCA CGC TTG CTC CCC GAT A	16S rRNA	*B. infantis*	28
PLO	TCA CGC TTG CTC CCC GAT A	16S rRNA	*B. longum*	28

REFERENCES

1. Charteris, W. P., Kelly, P. M., Morelli, L. and Collins, J. K. Selective detection, enumeration and identification of potentially probiotic *Lactobacillus* and *Bifidobacterium* species in mixed bacterial populations. *Int. J. Food Microbiol.* **35:** 1–27, 1997.
2. Cocconcelli, P. S., Porro, D., Galandini, S. and Senini, L. Development of RAPD protocol for typing of strains of lactic acid bacteria and enterococci. *Lett. Appl. Microbiol.* **21:** 376–379, 1995.

3. Delley, M., Mollet, B., and Hottinger, H. DNA probe for *Lactobacillus delbrueckii*. *Appl. Environ. Microbiol.* **56**(6): 1967–1970, 1990.
4. Drake, M., Small, C. L., Spence, K. D., and Swanson, B. G. Rapid detection and identification of *Lactobacillus* spp. in dairy products by using the polymerase chain reaction. *J. Food Prot.* **59**(10): 1031–1036, 1996.
5. Ehrmann, M., Ludwig, W., and Schleifer, K.-H. Species specific oligonucleotide probe for the identification of *Streptococcus thermophilus*. *Syst. Appl. Microbiol.* **15:** 453–455, 1992.
6. Ehrmann, M., Ludwig, W., and Schleifer, K.-H. Reverse dot blot hybridisation: A useful method for the direct identification of lactic acid bacteria in fermented food. *FEMS Microbiol. Lett.* **117**(2): 143–150, 1994.
7. Hensiek, R., Krupp, G., and Stackebrandt, E. Development of diagnostic oligonucleotide probes for four *Lactobacillus* species occurring in the intestinal tract. *Syst. Appl. Microbiol.* **15:** 123–128, 1992.
8. Hertel, C., Ludwig, W., Obst, M., Vogel, R. F., Hammes, W. P., and Schleifer, K.-H. 23S rRNA-targeted oligonucleotide probes for the rapid identification of meat *lactobacilli*. *Syst. Appl. Microbiol.* **14:** 173–177, 1991.
9. Johansson, M.-L., Molin, G., Pettersson, B., Uhlén, M., and Ahrné, S. Characterization and species recognition of *Lactobacillus plantarum* strains by restriction fragment length polymorphism (RELP) of the 16S rRNA gene. *J. Appl. Bacteriol.* **79:** 536–541, 1995.
10. Johansson, M.-L., Quednau, M., Molin, G., and Ahrné, S. Randomly amplified polymorphic DNA (RAPD) for rapid typing of *Lactobacillus plantarum* strains. *Lett. Appl. Microbiol.* **21:** 155–159, 1995.
11. Langendijk, P. S., Schut, F., Jansen, G. J., Raangs, G. C., Kamphuis, G. R., Wilkinson, H. F., and Welling, G. W. Quantitative fluorescence in situ hybridisation of *Bifidobacterium* spp. with genus-specific 16S rRNA-targeted probes and its application in faecal samples. *Appl. Environ. Microbiol.* **61**(8): 3069–3075, 1995.
12. Le Jeune, C., Lonvaud-Funel, A., ten Brink, B., Hofstra, H., and van der Vossen, J. M. B. M. Development of a detection system for histidine decarboxylating lactic acid bacteria based on DNA probes, PCR and activity test. *J. Appl. Bacteriol.* **78:** 316–326, 1995.
13. Lick, S., and Teuber, M. Construction of a species-specific DNA oligonucleotide probe for *Streptococcus thermophilus* on the basis of a chromosomal lacZ gene. *Syst. Appl. Microbiol.* **15:** 456–459, 1992.
14. Miteva, V. I., Abadjieva, A. N., and Stefanova, Tz. T. M13 DNA fingerprinting, a new tool for classification and identification of *Lactobacillus* spp. *J. Appl. Bacteriol.* **73:** 349–354, 1992.
15. Pébay, M., Colmin, C., Guédon, G., de Gaspéri, C., Decaris, B., and Simonet, J. M. Detection of intraspecific DNA polymorphism in *Streptococcus salivarius* subsp. *thermophilus* by a homologous rDNA probe. *Res. Microbiol.* **143:** 37–46, 1992.
16. Petrick, H. A. R., Ambrosio, R. E., and Holzapfel, W. H. Isolation of a DNA probe for *Lactobacillus curvatus*. *Appl. Environ. Microbiol.* **54:** 405–408, 1988.
17. Pilloud, N., and Mollet, B. DNA probes for detection of *Lactobacillus helveticus*. *Syst. Appl. Microbiol.* **13:** 345–349, 1990.
18. Pot, B., Hertel, C., Ludwig, W., Descheemaker, P., Kersters, K., and Schleifer, K.-H. Identification and classification of *Lactobacillus acidophilus, L. gasseri* and *L. johnsonii*

strains by SDS-PAGE and rRNA-targeted oligonucleotide probe hybridization. *J. Gen. Microbiol.* **139**(3): 513–517, 1993.

19. de Los Reyes-Gavilán, C. G., Limsowtin, G. K. Y., Tailliez, P., Séchaud, L., and Accolas, J.-P. A *Lactobacillus helveticus*-specific DNA probe detects restriction fragment length polymorphisms in this species. *Appl. Environ. Microbiol.* **58**(10): 3429–3432, 1992.
20. Rodtong, S., and Tannock, G. W. Differentiation of *Lactobacillus* strains by ribotyping. *Appl. Environ. Microbiol.* **59:** 3480–3484, 1993.
21. Rodtong, S., Dobbinson, S., Thode-Andersen, S., McConnell, M. A., and Tannock, G. W. Derivation of DNA probes for enumeration of a specific strain of *Lactobacilus acidphilus* in piglet digestive tract samples. *Appl. Environ. Microbiol.* **59**(11): 3871–3877, 1993.
22. Roussel, Y., Colmin, C., Simonet, J. M., and Decaris, B. Strain characterization, genome size and plasmid content in the *Lactobacillus acidophilus* group (Hansen and Mocquot). *J. Appl. Bacteriol.* **74:** 549–556, 1993.
23. Ståhl, M., Molin, G., Persson, A., Ahrné, S., and Ståhl, S. Restriction endonuclease patterns and multivariate analysis as a classification tool for *Lactobacillus* spp. *Int. J. Syst. Bacteriol.* **40**(2): 189–193, 1990.
24. Tannock, G. W. Biotin-labelled DNA probes for detection of epithelium-associated strains of *lactobacilli. Appl. Environ. Microbiol.* **55**(2): 461–464, 1989.
25. Tannock, G. W., McConnell, M. A., and Fuller, R. A note on the use of a plasmid as a DNA probe in the detection of a *Lactobacillus fermentum* strain in porcine stomach contents. *J. Appl. Bacteriol.* **73:** 60–62, 1992.
26. Vogel, R. F., Böcker, G., Stolz, P., Ehrmann, M., Fanta , D., Ludwig, W., Pot, B., Kersters, K., Schleifer, K.-H., and Hammes, W. P. Identification of lactobacilli from sourdough and description of *Lactobacillus pontis* sp. nov. *Int. J. Syst. Bacteriol.* **44**(2): 223–229, 1994.
27. Wang, R.-F., Cao, W.-W., and Cerniglia, C. E. PCR detection and quantitation of predominant anaerobic bacteria in human and animal fecal samples. *Appl. Environ. Microbiol.* **62**(4): 1242–1247, 1996.
28. Yamamoto, T., Morotomi, M., and Tanaka, R. Species specific oligonucleotide probes for five *Bifidobacterium* species detected in human intestinal flora. *Appl. Environ. Microbiol.* **58**(12): 4076–4079, 1992.

1.2.3 Commercial Probiotic Products

The following material was compiled from a survey conducted by Yakult Honsha, Japan, in 1997. The product information and functional claims were taken from the label on the bottle or package.

Japan

Product	Producer	Probiotics	Functional Claims	Product Category
Yakult 65	Yakult Honsha	*L. casei* strain Shirota	Normalizing the balance of human intestinal flora	Fermented milk drink
Yakult 80 Ace	Yakult Honsha	*L. casei* strain Shirota	Normalizing the balance of human intestinal flora	Fermented milk drink
Morinaga Caldus	Morinaga Milk Industry	*B. longum* *L. acidophilus*	Normalizing the balance of human intestinal flora	Fermented milk drink
Morinaga Bifidus	Morinaga Milk Industry	*B. longum* *L. acidophilus*	Normalizing the balance of human intestinal flora	Fermented milk
Sofuhl	Yakult Honsha	*L. casei* strain Shirota *S. thermophilus*	Normalizing the balance of human intestinal flora	Fermented milk
Joie	Yakult Honsha	*L. casei* strain Shirota *S. thermophilus*	Normalizing the balance of human intestinal flora	Fermented milk
Mil-Mil	Yakult Honsha	*B. breve* strain Yakult *B. bifidum* strain Yakult *L. acidophilus*	Normalizing the balance of human intestinal flora	Fermented milk
Mil-Mil E	Yakult Honsha	*B. breve* strain Yakult *L. acidophilus* *S. thermophilus*	Normalizing the balance of human intestinal flora	Fermented milk
Bifiel	Yakult Honsha	*B. breve* strain Yakult *S. thermophilus* *L. lactis* (*Lactococcus lactis*)	Normalizing the balance of human intestinal flora	Fermented milk
Levenin	Wakamoto Pharmaceutical	*L. acidophilus* *B. infantis* *E. faecalis*	Normalizing the balance of human intestinal flora	Freeze-dried powder
Onaka He GG	Takanashi Milk Products Co.	*S. thermophilus* *L. bulgaricus* *L. GG*	Normalize intestinal microflora	Fermented milk drink

(*continued*)

Asia, Oceania, and Brazil

Product	Producer	Probiotics	Functional Claims	Product Category	Country
Yakult	Yakult	*L. casei* strain Shirota	Normalizing the balance of human intestinal flora	Fermented milk drink	Taiwan, Brazil Hong Kong, Thailand, Korea, Philippines, Singapore, Mexico, Indonesia, Australia, Argentina
CHAMYTO	Nestlé	*L. johnsonii* *L. helveticus*	—	Fermented milk drink	Brazil
Vitagen	Malaysia Dairy Industry	*L. acidophilus*	Normalizing the balance of human intestinal flora	Fermented milk drink	Singapore, Malaysia
Bio-Garde	JALNA Dairy Food	*L. acidophilus* *B. bifidum* *L. casei*	—	Fermented milk	Australia
Bulla AB Live	Regal Cream Product	*L. acidophilus* *Bifidus*	—	Fermented milk	Singapore, Australia, Indonesia
Yoplait Yoplus	National Dairies	*L. acidophilus* *Bifidus*	—	Fermented milk	Singapore, Australia
VITA CHARM	P. T. Pola	*Lactobacilus* sp	Promoting health	Fermented milk	Indonesia
PAULS	QUF Industries Pauls	*L. acidophilus*	Promoting health	Fermented milk	Indonesia, Australia
VAALIA	QUF Industries Pauls	*L. acidophilus* *B. lactis* *L.* GG	Promotes health, balances intestinal microflora	Yogurt fermented milk smoothies	Australia
Ski	Australia Cooperation	*L. acidophilus* *Bifidus*	Promoting health	Fermented milk	Australia, Singapore
Classic Flavor	Danone	*L. acidophilus*	Promoting health	Fermented milk	Singapore
PAIGEN TWINU	CP-MEIJI	*L. bulgaricus* *L. acidophilus*	Promoting health	Fermented milk	Thailand

(*continued*)

Europe

Product	Producer	Probiotics	Functional Claims	Product Category	Country
Yakult	Yakult	*L. casei* strain Shirota	Normalizing the balance of human intestinal flora	Fermented milk drink	Netherlands, Belgium, United Kingdom, Germany
LC1	Nestlé	*S. thermophilus* *L. bulgaricus* *L. johnsonii*	Stimulating immune system	Fermented milk	France, Belgium, Spain, Switzer-land, Portugal, Italy, United Kingdom, Germany
Vifit	Campina	*S. thermophilus* *L. bulgaricus* *L. acidophilus* *B. bifidum* *L. casei* GG	Keeping the intestinal flora in good shape, enhancing immune response	Fermented milk	Netherlands, Belgium, United Kingdom
Vifit	Südmilch	*S. thermophilus* *L. bulgaricus* *L. acidophilus* *B. bifidum* *L. casei* GG	Keeping the intestinal flora in good shape, enhancing immune response	Fermented milk	Germany
GEFILUS	Valio Ltd	*S. thermophilus* *L. bulgaricus* *L. acidophilus* *B. bifidum* *L. casei* GG	Keeping the intestinal flora in good shape, enhancing immune response	Fermented milk	Finland
Actimel	Danone	*S. thermophilus* *L. bulgaricus* *L. casei* Imuntass	Enhancing resistance to harmful bacteria	Fermented milk	Netherlands, Belgium
Fyos	Nutricia	*L. casei* subsp. *rhamnosus* NCC 208	—	Fermented milk	Netherlands, Belgium
LA7	Bauer	*L. acidophilus* LA7 Bifidus	Improving intestinal flora	Fermented milk	Germany
Bio Pot	Onken Dairy	*L. acidophilus* *B. bifidum*	Improving intestinal flora	Fermented milk	Germany

(*continued*)

Product	Producer	Probiotics	Functional Claims	Product Category	Country
Ventrux Acido	A B Cernelle	*S. feacium* SF 68	Nutritional supplement	Capsule	Sweden
Bra-Mjoelk	Arla	*L. reuteri*	Defense system for bowel	Fermented milk	Sweden
Biola	Tine	*B. lactis* *L. acidophilus* *L.* GG	Balance microflora	Yogurt, fermented milk	Norway
Little Swallow	Schwalbachen-AG	*L. acidophilus* *B. bifidum*	—	Fermented milk	United Kingdom, Germany
Leisure Live	Leisure Fine Foods & Drinks	*L. acidophilus* *B. bifidum*	—	Fermented milk	United Kingdom
Sym-Balance	Toni AG	*L. reuteri* *L. acidophilus* *L. casei* *Bifidobacteria*	Defense system for bowel	Fermented milk	Switzerland
Aktifit	Emmi AG	*S. thermophilus* *L. casei* *L. acidophilus* *B. bifidum*	Defense system for bowel	Fermented milk	Switzerland
Bioforin	Giuliani S.A.	*S. feacium* SF 68	Nutritional supplement	Capsule	Switzerland
LC1	Nestlé	*S. thermophilus* *L. bulgaricus* *L. acidophilus* NCC 208	Stimulating immune system	Capsule	France, Finland Belgium, Spain, Switzerland, Portugal, United Kingdom, Germany
Lactophilus	Laboratoires Lyocentre	*L. casei* subsp. *rhamnosus*	—	Freeze-dried powder	France

United States of America

Product	Producer	Probiotics	Functional Claims	Product Category
Erivan Acidophilus Yoghurt	Erivan Dairy	*L. acidophilus*	Promoting health	Fermented milk
Crunch N-Yoghurt	Yoplait	*S. thermophilus* *L. bulgaricus* *L. acidophilus*	Promoting health	Fermented milk
Classic Flavor	Danone	*L. acidophilus*	Promoting health	Fermented milk

(*continued*)

Product	Producer	Probiotics	Functional Claims	Product Category
Lactinex	Hynson, Westcott and Dunning	*L.acidophilus* *L. bulgaricus*	Promoting health	Powder
LGG	ConAgra	*L.* GG	Promoting health	Capsules

1.3 SAFETY OF PROBIOTIC BACTERIA

There is considerable interest in extending the range of foods containing probotic organisms from dairy foods to infant formulae, baby foods, fruit juice-based products, cereal-based products, and pharmaceuticals. New species and more specific strains of probiotic bacteria are constantly being sought. Traditional probiotic dairy strains of lactic acid bacteria have a long history of safe use, and most strains are considered commensal microorganisms with no pathogenic potential. As yet, no general guidelines exist for safety testing of probiotics. However, some recommendations are given in the review of Donohue and Salminen (6) and Donohue et al. (5, 6). Different aspects of the safety of probiotic bacteria should be assessed using a panel of in vitro methods, animal models, and human subjects.

In the food industry the most widely used probiotic bacteria belong to the group of lactic acid bacteria, though some bifidobacteria and yeasts are also utilized. The term "lactic acid bacteria" currently includes the genera *Lactobacillus, Leuconostoc, Pediococcus,* and *Lactococcus.* Although some strains of *Streptococcus* and *Enterococcus* share the properties of lactic acid bacteria, *Streptococcus thermophilus* is the only strain currently used in fermented dairy products.

Safety of viable microorganisms is difficult to establish using the current assessment methods. Theoretically, almost any microorganism may under certain circumstances either cause an infection or alter its virulence due to internal or external factors. Our gastrointestinal tract harbors around 500 known species of bacteria with a great number of strains in each species. The balance between the intestinal microbes, the intestinal barrier, and our body forms a cohabitat in which the human system accepts the microbial interference and ecology in its current form. Some microbes have virulence factors that make them pathogenic and infectious. It may be the dose or concentration of such microbes locally in the gastrointestinal tract and the physiological and immunological state of the host that may transform others, such as enterococci, from harmless commensals to opportunistic pathogens.

The use of lactic acid bacteria in foods has a long history, and most strains are considered commensal microorganisms with no pathogenic potential. Their ubiquitous presence in intestinal epithelium and the human gastrointestinal tract, and their traditional use in fermented foods and dairy products without significant problems, attest to their safety. Members of the genus *Lactobacillus* are most commonly given safe or generally recognized as safe (GRAS) status, while members of the genera *Streptococcus* and *Enterococcus* contain many opportunistic pathogens.

The safety of probiotics has been questioned in recent reviews and clinical reports, which have drawn attention to cases of human bacteraemia associated with the presence of lactic acid bacteria (1, 2, 7).

A variety of strains of probiotic organisms have been used in the clinical treatment of gastrointestinal disorders in both children and adults. These include conditions in which mucosal integrity is impaired by antibiotics or radiotherapy, or acute diarrhea as a result of bacterial or viral origin; the organisms are also used in prevention of gut colonization by pathogens (13). No evidence of opportunistic infections or other ill effects by probiotics have been observed in these studies. Also, animal studies indicate an absence of infectivity, and specific toxicity studies show no signs of toxic or harmful effects even at extremely high dose levels (6, 11, 14).

Degradation of intestinal mucus has also been used as a marker of toxicity. It is thought that a stable gastrointestinal microflora with normal patterns of fermentation and colonization resistance and low pH are important in protecting the mucosal layer from injury (16). Strains that do not degrade intestinal mucus or its glycoproteins are thought to be noninvasive. Strains that do not degrade intestinal mucosa are also thought to be therapeutic in the probiotic treatment of mucosal diseases such as pouchitis, ulcerative colitis, and Crohn's disease. In a recent study, commercial probiotic strains (*Lactobacillus* GG, *Lactobacillus acidophilus, Bifidobacterium bifidum*) were shown to be inactive in mucosal degradation (16). In earlier studies, some fecal Bifidobacteria were found to take part in mucus degradation (9, 16).

Recently, the association of lactic acid bacteria in germ-free rodents has also been used as a criterion for safety. Ruseler van Embden et al. (16) studied the association or colonization of germ-free rodents with several probiotic lactobacilli and detected no adverse effects in these animals.

Clinical Studies

A large amount of data from clinical trials or studies in human volunteers also attest to the safety of lactic acid bacteria. These studies have included short-term trials in normal volunteers; prevention and treatment of acute diarrhea in premature infants, infants, and children; studies on immune effects; and studies in patients with severe intestinal infections. A study using *Lactobacillus acidophilus* preparations in the effective prevention of intestinal side effects during pelvic radiotherapy has also been reported. Aso et al. reported that the recurrence-free interval after resection of superficial bladder cancer in humans was extended by treatment with *Lactobacillus casei* Shirota strain. Available data indicate that no harmful effects have been observed in controlled clinical studies with lactobacilli and bifidobacteria. To the contrary, during treatment of intestinal infections beneficial effects have been observed including stabilization of gut mucosal barrier, prevention of diarrhea, and amelioration of infant and antibiotic-associated diarrhea.

Table 1.3 Examples of Potential Studies and Safety Effects with Probiotic Bacteria (20, 21, 22)

Type of Study	Potential Effects
in vitro Studies	
Invasiveness	Adhesion but no invasion, prevention of pathogen invasion, competitive exclusion
Mucus degradation	No mucus degradation observed
Antimicrobial production	Prevention of pathogen growth
Animal studies	
DMH induced tumor formation	Tumor development delayed in animals
Alcohol induced liver damage	Prevention or alleviation of liver damage in experimental alcoholic liver disease
Clinical studies	
Clinical studies	Prevention or treatment of diseases or dysfunctions, side effects or harmful effects
Colonization of pre-term infants, infants, and children	Colonization and potential side effects
Epidemiological studies/Post-marketing surveillance	
Infection potential/genetic methods	Probiotic related infections No lactic acid bacteria related infections in normal or immune compromised subjects

Post-marketing Epidemiological Surveillance

Case reports from the literature of lactic acid bacteria in association with clinical infection in humans have recently been analyzed in reviews by Gasser (7) with Aguirre and Collins (2). Both reviews conclude that considering their widespread consumption lactic acid bacteria appear to have very low pathogenic potential. Two recent Finnish studies confirm that the number of infections associated with lactic acid bacteria is extremely small and no cases could be linked to commercial probiotics (23, 24).

Safety of Novel Probiotics

There is considerable interest in extending the range of foods incorporating probiotic organisms from dairy foods to infant formulae, baby foods, fruit juice-based products, cereal-based products, and pharmaceuticals (13). New species and more specific strains of probiotic bacteria are being sought. It cannot be assumed that these novel probiotic organisms share the historical safety of traditional strains. Thus, the European Union novel foods regulations allow a set approach for safety assessment of such strains X, Y. Before their incorporation into products, new strains should be carefully assessed and tested for the safety and efficacy of their proposed use. Especially the following areas should be studied:

a) Assessment of the intrinsic properties of the probiotic using, for example, adhesion factors, antibiotic resistance (plasmid profile, plasmid transfer potential, transfer of vancomycin resistance to Enterococci), and enzyme profile in comparison to food and intestinal strains.
b) Assessment of the safety and effects of the metabolic products of the probiotic.
c) Assessment of the acute toxicity of ingestion of extremely large amounts of the probiotic.
d) Estimation of the in vitro invasiveness of the probiotic using cell lines and human intestinal mucus degradation. Assess infectivity in animal models including gnotobiotic or lethally irradiated animals.
e) Determination of the efficacy of ingested probiotic as measured by dose-response (minimum and maximum dose required, consequent health effects); assess the effect of massive probiotic doses on the composition of human intestinal microflora.
f) Careful assessment with no side effects during human volunteer studies and clinical studies in various disease-specific states.
g) Epidemiological surveillance of people ingesting large amounts of the probiotic or fermented food products.
h) Assessment of the "novel" status in EU regulations.

REFERENCES

1. Adams, M. R., and Marteau, P. On the safety of lactic acid bacteria from food. *Int. J. Food Microbiol.* **27:** 263–264, 1995.
2. Aguirre, M., and Collins, K. Lactic acid bacteria and human clinical infection. *J. Appl. Bacteriol.* **75:** 95–107, 1993.
3. Aso, Y., and Akazan, H. Prophylactic effect of a *Lactobacillus casei* preparation on the recurrence of superficial bladder cancer. *Urol. Int.* **49:** 125–129, 1992.
4. Coconnier, M. H., Klaenhammer, T. R., Kerneis, S., Bernet, M. F., and Servin, A. Protein mediated adhesion of *Lactobacillus acidophilus* BG2F04 on human enterocyte and mucus secreting cell lines in culture. *Appl. Environ. Microbiol.* **58:** 2034–2039, 1992.
5. Donohue, D., Deighton, M., Ahokas, J. T., and Salminen, S. Toxicity of lactic acid bacteria. In: *Lactic Acid Bacteria.* S. Salminen and A. von Wright (Eds.). Marcel Dekker Inc., New York, 1993, pp. 307–313.
6. Donohue, D. C., and Salminen, S. Safety of probiotic bacteria. *Asia Pac. J. Clin.* **5:** 25–28, 1996.
7. Gasser, F. Safety of lactic acid bacteria and their occurrence in human clinical infections. *Bull. Inst. Pasteur* **92:** 45–67, 1994.

8. Harty, D. W. S., Oakey, H. J., Patrikaki, M., Hume, B. B. H., and Knox, K. W. Pathogenic potential of lactobacilli. *Int. J. Food Microbiol.* **24:** 179–189, 1994.
9. Hoskins, L. C., Augustines, M., McKee, W. B., Boulding, E. T., Kriaris, M., and Niedermeyer, G. Mucin degradation in human colon ecosystems. Isolation and properties of faecal strains that degrade ABH blood group antigens and oligosaccharides from mucin glycoproteins. *J. Clin. Invest.* **75:** 944–953, 1985.
10. ILSI Europe. A scientific basis for regulations on pathogenic micro-organisms in foods. Summary of a workshop held in May 1993 and organized by the Scientific Committee on Microbiology. ILSI Press, 1993.
11. Ishihara, K., Miyakawa, H., Hasegawa, A., Takazoe, I., and Kawai, Y. Growth inhibition of *Streptococcus mutans* by cellular extracts of human intestinal lactic acid bacteria. *Infect. Immun.* **3:** 692–694, 1985.
12. Jonas, D. A., Antignac, E., Antoine, J. M., Classen, H. G., Huggett, A., Knudsen, I., Mahler, J., Ockhuizen, T., Smith, M., Teuber, M., Walker, R., and De-Vogel, P. The safety assessment of novel foods. Guidelines prepared by ILSI Europe Novel Food Task Force. *Food Chem. Toxicol.* **34**(10): 931–940, 1996.
13. Lee, Y. K., and Salminen, S. The coming of age of probiotics. *TIFST.* **6:** 241–245, 1995.
14. Momose, H., Igarashi, M., Era, T., Fukuda, Y., Yamada, M., and Ogasa, K. Toxicological studies on *Bifidobacterium longum* BB-536. *Pharmacometrics* **17:** 881–887, 1979.
15. Oakey, H. J., Harty, D. W. S., and Knox, K. W. Enzyme production by lactobacilli and the potential link with infective endocarditis. *J. Appl. Bacteriol.* **78:** 142–148, 1995.
16. Ruseler-van Embden, J. G. H., Liesholt, L. M. C., Gosselink, M. J., and Marteau, P. Inability of *Lactobacillus casei* strain GG, *L. acidophilus* and *Bifidobacterium bifidum* to degrade intestinal mucus glycoproteins. *Scand. J. Gastroenterol.* **30:** 675–680, 1995.
17. Saavedra, J. M. Microbes to fight microbes: A not so novel approach to controlling disease. *J. Pediatr. Gastroenterol. Nutr.* **21:** 125–129, 1995.
18. Salminen, E., Elomaa, I., Minkkinen, J., Vapaatalo, H., and Salminen, S. Preservation of intestinal integrity during radiotherapy using live *Lactobacillus acidophilus* cultures. *Clin. Radiol.* **39:** 435–437, 1988.
19. Salminen, E., Salminen, S., and Vapaatalo, H. Adverse effects of pelvic radiotherapy. *Progress in Radio-Oncology. Vienna.* 501–504, 1995.
20. Salminen, S., Bouley, M. C., Boutron-Rualt, M. C., Cummings, J., Franck, A., Gibson, G., Isolauri, E., Moreau, M. C., Roberfroid, M., and Rowland, I. Gastrointestinal physiology and function; the role of prebiotics and probiotics. *Br. J. Nutr.* **80,** Suppl. 1: 147–151, 1998a.
21. Salminen, S., and von Wright, A. Current human probiotics—safety assured? *Microbial. Ecol. Health Dis.* 1998 (in press).
22. Salminen, S., von Wright, A., Morelli, L., Marteau, P., Brassard, D., de Vos, W., Fondén, R., Saxelin, M., Collins, K., Mogensen, G., Birkeland, S. E., and Mattila-Sandholm, T. Demonstration of safety of probiotics—a review. *Int. J. Food Microbiol.* 1998b (in press).
23. Saxelin, M., Rautelin, H., Chassy, B., Gorbach, S. L., Salminen, S., and Mäkelä, P. Lactobacelli and septic infections in Southern Finland during 1989–1992. *Clin. Infect. Dis.* **22:** 564–566, 1996.
24. Saxelin, M., Rautelin, H., Salminen, S., and Mäkelä, P. The safety of commercial products with viable *Lactobacillus strains. Infect. Dis. Clin. Pract.* **5:** 331–335, 1996.

1.4 LEGAL STATUS OF PROBIOTICS

There has been much debate on whether existing legislatory control is sufficient for the regulation of functional foods and whether specific registration has to be introduced. At this moment, legal definition of a functional food has been stipulated under the Food for Specific Health Use (FOSHU) system of Japanese food regulation. A functional food is defined as a food, that claims to have a positive effect on health.

	Standard for Functional Food	Position on Health Claims
Codex Alimentarius	No regulation	Not permitted
Japan	FOSHU	Permitted under FOSHU rule
United States	No regulation	Permitted
Canada	No regulation	Permitted/under review
Australia/New Zealand	No regulation	Not permitted/under review
European Union	No regulation	Not permitted
Sweden	No regulation	Permitted specific claims (as agreed by a special group of experts and industry)
Netherlands	No regulation	Self regulated specific claims approved by a group of experts

2

SELECTION AND MAINTENANCE OF PROBIOTIC STRAINS

2.1 ISOLATION OF PROBIOTIC MICROORGANISMS

Various probiotic microorganisms could be isolated from the mouth, gastrointestinal content, and feces of animal and human, by repetitive subculturing of the microorganisms on appropriate enrichment or selective media as listed below.

Isolation of Probiotic Microorganisms[a,b]

Organisms	Nonselective Medium	Selective Medium	Supplements	Culture Conditions
Lactobacilli	MRS medium	SL medium	Cycloheximide,[c] (100 mg/l) Cystein (0.05 % w/v)[d] Growth factors[e] Other carbohydrates[f]	10% CO_2 + 90 N_2 or H_2
Bifidobacteria	MRS medium	TPY medium[g]	Antibiotics[h]	37–40°C in anaerobic jar, 10% CO_2 + 90% H_2
Streptococci	Strep Base medium[i]	TYC medium[k]	—	5% CO_2 in air
Enterococci	Brain–heart infusion medium[j]	Kanamycin aesculin medium[l]	—	>3% CO_2 in air

[a]After Butler, J. P. (ed.), *Bergey's Manual of Systematic Bacteriology,* Vol. 2. Williams & Wilkins, Baltimore, 1986.

(*continued*)

[b]After Balows, A., Trüper, H., Devorkin, M., Harder, W., and Schleifer, K.-H. (eds.), *The Prokaryotes. A Handbook on the Biology of Bacteria: Ecophysiology, Isolation, Identification, Applications.* Vol. 2. Springer-Verlag, New York, 1992.
[c]To eliminate yeasts.
[d]To isolate anaerobic lactobacilli, from intestinal sources.
[e]Examples are meat extract, tomato juice, fresh yeast extract, malt extract, ethanol, mevalonic acid (sake), beer and juices, to improve the isolation of lactobacilli, which are adapted to a particular ecological niche.
[f]Replacement of glucose by maltose, fructose, sucrose, or arabinose is recommended for the isolation of heterofermentative lactobacilli.
[g]TPY medium: trypticase (BBL), 10 g; phytone (BBL), 5 g; glucose, 5 g; yeast extract (Difco), 2.5 g; Tween 80, 1 ml; cysteine hydrochloride, 0.5 g; K_2HPO_4, 2 g; $MgCl_2 \cdot 6H_2O$, 0.5 g; $ZnSO_4 \cdot 7H2O$, 0.25 g; $CaCl_2$, 0.15 g; $FeCl_3$, a trace; agar, 15 g; distilled water to 1000 ml. Final pH is about 6.5 after autoclaving at 121°C for 25 min; dilutions can be made with the same liquid medium.
[h]Use of kanamycin, neomycin, paramomycin, sodium propionate, lithium chloride, sorbic acid, or sodium azide could improve selectivity in some cases.
[i]Strep Base: Proteose peptone, 20 g/l; yeast extract, 5 g/l; NaCl, 5 g/l; Na_2HPO_4, 1 g/l, glucose, 5 g/l. Dissolve ingredients in distilled water, adjust pH to 7.6. Autoclave at 121°C for 15 min.
[j]Brain–heart infusion medium: Calf brain infusion solids, 12.5 g/l; beef heart infusion solids, 5.0 g/l; proteose peptone, 10.0 g/l; dextrose, 2.0 g/l; NaCl, 5.0 g/l; Na_2HPO_4, 2.5 g/l. pH 7.4. Sterilize by autoclaving at 121°C for 15 min.
[k]TYC medium: Trypticase, 15.0 g/l; yeast extract, 5.0 g/l; L-cystine, 0.2 g/l; Na_2SO_3, 0.1 g/l; NaCl, 1.0 g/l; $NaHCO_3$, 2.0 g/l; Na_2HPO_4 $12H_2O$, 2.0 g/l; sodium acetate ($3H_2O$), 20.0 g/l; sucrose, 50.0 g/l. Adjust pH to 7.3. Autoclave at 121°C for 15 min.
[l]Kanamycin aesculin medium: Tryptone, 20 g/l; yeast extract, 5 g/l; NaCl, 5 g/l; sodium citrate, 1 g/l; aesculin, 1 g/l; ferric ammonium citrate, 0.5 g/l; odium azide, 0.15 g/l; Kanamycin sulfate, 0.02 g/l. Bring to boil to dissolve completely. Sterilize by autoclaving at 121°C for 15 min.

2.2 SELECTION OF PROBIOTIC MICROORGANISMS

Common criteria used for isolating and defining probiotic bacteria and specific strains include the following: genera of human origin, bile and acid stability, adhesion to intestinal mucosa, temporary colonization of the human gastrointestinal tract, production of antimicrobial components, and safety in human use (7, 8). The development of new probiotics has been based on the outlined decision tree presented in Figure 2.1. Most current probiotics have been selected in a similar manner. However, the outcome of such selection has also been questioned. The most often used genera are lactobacilli and bifidobacteria. They can be given with fermented foods such as yogurt or fermented vegetables or meats, and they may briefly established in the gut. The following areas have been of special interest recently.

Immunological Assessment. Gut-associated lymphoid tissue can have a long contact with adhesive probiotic preparations, and adhesion is one way of provoking immune effects. Immune activation has been suggested to be responsible for prevention and stabilization of acute gastroenteritis in humans. Immune effects of probiotics have also been related to probiotic effects seen in colon and bladder cancer studies. Thus, beneficial immune responses need to be further studied for future probiotic strains and in vitro studies correlated with human clinical trials (see Section 5.3).

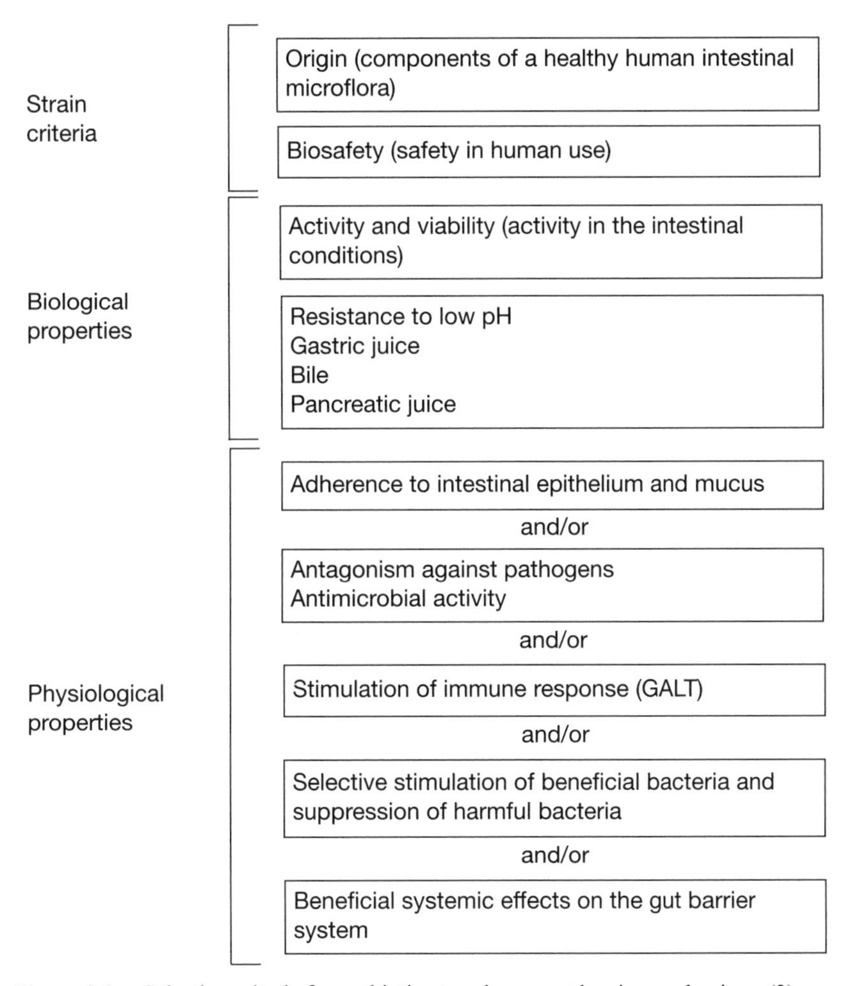

Figure 2.1. Selection criteria for probiotics to enhance gut barrier mechanisms (2).

Production of Antibacterial Substances. Lactic acid bacteria commonly produce a wide variety of antibacterial substances including bacteriocins, bacteriocin-like substances, antibiotics, as well as lactic acid and hydrogen peroxide. These substances promote successful colonization by improving the competitive advantage of the introduced strain against the established normal inhabitants of the gastrointestinal tract (7, 8) (see Section 5.2).

Adhesive Properties. Adhesion of probiotics to intestinal cells is important for some applications. Thus, novel selection criteria include the use of at least two different methods for studying adhesion of probiotics on human intestinal cell lines or human ileal or intestinal mucus preparations (3, 5, 6, 12).

Technological Properties. It is also of importance that the technological properties make probiotic lactic acid bacteria easily accessible to the food industry. Hardy growth and pleasant aroma and flavor profiles are of importance when probiotic functional foods are developed. Also the stability of probiotic strains in continuous industrial culturing and functional foods provides challenges for the industry (4).

Novel Selection Criteria. These include among others immunological assessment using peripheral blood lymphocytes or activation of phagocytosis as measures of immune enhancement or suppression. It is likely significant new developments will occur in this area in the future (7).

Safety. The safety of lactic acid bacteria used in clinical and functional food is of great importance. In general, lactic acid bacteria have a good record of safety, and no major problems have occurred. Cases of infection have been reported with several strains, most commonly with those naturally most abundant in the human intestinal mucosa. Safety has been documented with dairy strains (1, 10, 11), and a review of current knowledge on safety of probiotic bacteria has been published by Donohue and Salminen and the EU demonstration group (1, 9). It is most important for future probiotic lactic acid bacteria that their safety has been assured and that they conform to all regulations.

REFERENCES

1. Donohue, D. C., and Salminen, S. Safety of probiotic bacteria. *Asia–Pac. J. Clin. Nutr.* **5:** 25–28, 1996.
2. Huis In't Veld, J. H. J,. and Shortt, C. Selection criteria for probiotic microorganisms. The Royal Society of Medicine, International Congress and Symposium Series 1996, **219:** 27–36, 1996.
3. Kirjavainen, P. V., Ouwehand, A. C., Isolauri, E., and Salminen, S. J. The ability of probiotic bacteria to bind to human intestinal mucus. *FEMS Microb. Lett.* **167:** 185–189, 1998.
4. Lee, Y. K., and Salminen, S. The coming of age of probiotics. *Trends Food Sci. Technol.* **6:** 241–245, 1995.
5. Lehto, E., and Salminen, S. Adhesion of two *Lactobacillus* strains, ond *Lactococcus* strain and one *Propionibacterium* strain to cultured human intestinal Caco-2 cell line. *Bioscience and Microflora* **16:** 13–17, 1997.
6. Ouwehand, A. C., Isolauri, E., Kirjavainen, P. V., and Salminen, S. J. Adhesion of four *Bifidobacterium* strains to human intestinal mucus from subjects in different age groups. *Letters in App.l Microbiol.* 1998 (in press).
7. Salminen, S., Deighton, M., Benno, Y., and Gorbach, S. L. Lactic acid bacteria in health and disease. In *Lactic Acid Bacteria; Microbiology and Functional Aspects* (S. Salminen and A. von Wright, eds.), pp. 211–253, Marcel Dekker, Inc., New York. 1998.
8. Salminen, S., Isolauri, E., and Salminen, E. Clinical uses of probiotics for stabilizing the gut mucosal barrier: successful stains and future challenges. *Antonie van Leeuwenhoek.* **70:** 347–358, 1996.

9. Salminen, S., von Wright, A., Morelli, L., Marteau, P., Brassart, D., de Vos, W., Fonden, R., Saxelin, M., Coleins, K., Mogeuseen, G., Berkeland, S. G., and Maltila-Sandholm, T. Demonstration of safety of probiotics—a review. *Int. J. Food Microbiol.* 1998 (in press).
10. Saxelin, M., Chuang, N. H., Chassy, B., Rautelin, H., Mäkelä, H., Salminen, S., and Gorbach, S. L. Lactobacilli and bacteremia in Southern Finland, 1989–1992. *Clin. Infect. Dis.* **22:** 564–566, 1996.
11. Saxelin, M., Rautelin, H., Salminen, S., and Mäkelä, H. The safety of commercial products with viable *Lactobacillus* strains. *Infect. Dis. Clin. Pract.* **5:** 331–335, 1996.
12. Tuomola, E., and Salminen, S. Adhesion of some probiotic and dairy *Lactobacillus* strains to Caco-2 cell cultures. *Int. J. Food Microbiol.* **41:** 45–51, 1998.

2.2.1 Origin of Isolates

Most of the lactobacilli isolated from a specific site of a specific animal source could only colonize epithelium in the same kind; some lactobacilli isolated from gastrointestinal tract could not colonize all the animals tested.

Adhesion of Lactobacilli to Squamous Epithelial Cells[a]

	Adhesion to Epithelial Cells of				
Animal Source	Mouse	Rat	Chicken	Pig	Ref.
Mouse stomach	9/9	1/1	0/5	ND	6
Rat stomach	4/4	2/2	0/5	0/1	8
Rat stomach	0/4	0/1	ND	ND	8
Rat stomach	0/4	0/2	ND	ND	4
Fowl crop	0/8	0/4	7/7	0/1	2
Fowl crop	0/5	0/1	0/7	ND	2
Fowl crop	0/4	0/1	0/5	ND	8
Fowl crop	ND	ND	5/5	ND	3
Fowl crop	ND	ND	ND	4/6	1
Pig stomach	0/5	0/1	0/5	0/1	8
Pig intestine	ND	ND	ND	15/17	1

[a]Denominators indicate total number of animals tested; ND, not done.

2.2.2 Stability During Storage

Cell Viability During Cold Storage of Cultured Milk at 4°C[a]

		Viable Cell Concentration (% initial)			
	Initial Cell Concentration	Days			
Organisms	(CFU/ml)	0	14	28	49
Lactobacillus casei strain Shirota	1.6×10^8	100	93.8	93.8	87.5
L. casei NCIMB 8822	1.3×10^8	100	43.1	6.8	2.5
L. casei NCIMB 11970	1.5×10^8	100	43.3	8.0	4.6

(continued)

Cell Viability During Cold Storage of Cultured Milk at 4°C[a] *(cont.)*

Organisms	Initial Cell Concentration (CFU/ml)	Viable Cell Concentration (% initial) Days 0	14	28	49
L. paracasei NCIMB 8001	3.6×10^8	100	100	94.4	83.3
L. paracasei NCIMB 9709	1.4×10^8	100	100	100	100
L. paracasei NCIMB 9713	3.3×10^8	100	100	75.8	42.4
L. rhamnosus NCIMB 6375	4.8×10^8	100	64.5	10.4	4.0
L. rhamnosus GG	2.0×10^8	100	90	50	49
L. acidophilus NCIMB 8690	1.7×10^8	100	25.9	1.9	0.008
L. acidophilus CH 5	7.0×10^7	100	0.03	0.0014	0.001
L. bulgaricus NCIMB 11778	5.2×10^7	100	67.3	1.2	0.06
Streptococcus thermophilus NCIMB 10387	3.2×10^6	100	10.3	0.25	0.072

[a]Data provided by Y. K. Lee, National University of Singapore.

Cell Viability During Storage of Cultured Milk at 25°C[a]

Organisms	Initial Cell Concentration (CFU/ml)	Viable Cell Concentration (% initial) Days 0	14	28	49
L. casei strain Shirota	1.6×10^8	100	87.5	62.5	57.5
L. casei NCIMB 8822	1.3×10^8	100	60.8	57.7	55.4
L. casei NCIMB 11970	1.5×10^8	100	66.7	42	10
L. paracasei NCIMB 8001	3.6×10^8	100	72.2	20.8	3.9
L. paracasei NCIMB 9709	1.4×10^8	100	71.4	12.9	4.1
L. paracasei NCIMB 9713	3.3×10^8	100	93.9	17.9	4.
L. rhamnosus NCIMB 6375	4.8×10^8	100	25	7.1	1.9
L. rhamnosus GG	2.0×10^8	100	65	46.5	16
L. acidophilus NCIMB 8690	1.7×10^8	100	2.3	0.03	0
L. acidophilus CH 5	7.0×10^7	100	0.005	0	0
L. bulgaricus NCIMB 11778	5.2×10^7	100	21	0.002	0
S. thermophilus NCIMB 10387	3.2×10^6	100	0.013	0	0

[a]Data provided by Y. K. Lee, National University of Singapore.

2.2.3 Tolerance to Additives

The following data show the effect of natural fruit juice added on the time taken for the viable cell number of *L. acidophilus* CH5 (Chr. Hanson's Laboratory) to decrease by 2 log cycle (t_d) in cultured milk exposed to light. Storage temperature: 5°C, pH 3.8.

	t_d (days)
Milk	15
Milk + 3% strawberry juice (v/v)	5
Milk + 3–10% apple juice (v/v)	15
Milk + 3–10% grape juice (v/v)	15
Milk + 3–10% mango juice (v/v)	15
Milk + 3–10% orange juice (v/v)	15
Milk + 3–10% pineapple juice (v/v)	15

After Lee and Wong (5).

2.2.4 Stability During Passage to Intestinal Sites

The following data show the stability of lactic acid bacteria in synthetic gastric juice (0.2% bile salt w/v in MRS, pH 5.6) at 37°C

	Viable Cell Concentration (CFU[a]/ml)[b]	
	Initial	After 3 h
L. casei strain Shirota	2.0×10^8	1.0×10^8
L. casei NCIMB 8822	2.9×10^8	4.0×10^2
L. casei NCIMB 11970	5.8×10^8	5.7×10^2
L. paracasei NCIMB 8001	8.7×10^8	1.2×10^7
L. paracasei NCIMB 9709	3.2×10^8	0
L. paracasei NCIMB 9713	5.2×10^8	1.1×10^7
L. rhamnosus NCIMB 6375	4.5×10^8	6.9×10^6
L. rhamnosus GG	2.3×10^8	1.0×10^8
L. acidophilus NCIMB 8690	1.5×10^7	5.4×10^4
L. acidophilus CH 5	2.0×10^7	2.1×10^5
L. bulgaricus NCIMB 11778	9.6×10^6	0
S. thermophilus NCIMB 10387	2.6×10^6	6.1×10^3

[a]CFU, colony-forming unit.
[b]Data provided by Y. K. Lee, National University of Singapore.

2.2.5 Minimum Effective Dose

The following data [abstracted from Saxelin et al. (7)] show fecal recovery of lactobacilli (CFU/g feces) from healthy human volunteers (n = 20, 20–55 years) after oral administration of *L. rhamnosus* GG at dose levels of 1.6×10^8 (low) or 1.2×10^{10} (high) CFU/g in gelatin capsules for 7 days.

	Low Dose		High Dose	
Day	Lactobacilli[a]	*L.* GG[a]	Lactobacilli	*L.* GG
0	1.5×10^8 (10^5–1.5×10^8)	ND[b]	9.2×10^7 (10^5–3.4×10^8)	ND
3	1.7×10^7 (10^7–2.6×10^7)	ND	2.1×10^8 (1.8–49.0×10^7)	2.0×10^6 (10^3–5.6×10^6)
5	2.8×10^8 (1.5×10^6–3.8×10^8)	ND	2.5×10^8 (10^6–10^9)	1.5×10^6 (10^4–10^7)
7	8.0×10^7 (10^6–2.1×10^7)	10^3 (ND–10^3)[c]	9.8×10^7 (5.3×10^5–2.9×10^8)	1.2×10^5 (10^3–2.9×10^5)

[a]Colony-forming units determined on MRS agar plates, pH 6.2, 37°C, anaerobically for 3 days. *L. rhamnosus* GG was identified by the characteristic of colonies confirmed by Gram staining and sugar fermentation tests.
[b]Not detected.
[c]Detected only in one sample.

Remarks The lowest dose needed for fecal detection of *L. rhamnosus* GG was 10^{10} CFU/d as freeze-dried powder. *L. rhamnosus* GG did not significantly change the total number of fecal lactobacilli.

REFERENCES

1. Barrow, P. A., Brooker, B. E., Fuller, R., and Newport, M. J. The attachment of bacteria to the gastric epithelium of the pig and its importance in the microecology of the intestine. *J. Appl. Bacteriol.* **48:** 147–154, 1980.
2. Champ, M., Szylit, O., and Gallant, D. J. The influence of microflora on the breakdown of maize starch granules in the digestive tract of chicken. *Poult. Sci.* **60:** 179–187, 1981.
3. Fuller, R. Ecological studies on the *Lactobacillus* flora associated with the crop epithelium of the fowl. *J. Appl. Bacteriol.* **36:** 131–139, 1973.
4. Hemme, D., Raibaud, P., Ducluzeau, R., Galpin, J.-V., Sicard, P., and van Heijenoort, J. *Lactobacillus murinus* n. sp., une nouvelle espèce de la flore dominate autochtone du tube digestif du rat et de la souris. *Ann. Microbiol. (Paris)* **131A:** 297–308, 1980.
5. Lee, Y. K., and Wong, S. F. Stability of lactic acid bacteria in fermented milk. In *Lactic Acid Bacteria* (S. Salminen and A. von Wright, eds.), 2nd ed. Dekker, New York, 1998, pp. 103–114.
6. Roach, S., Savage, D. C., and Tannock, G. W. *Lactobacilli* isolated from the stomach of conventional mice. *Appl. Environ. Microbiol.* **33:** 1197–1203, 1977.
7. Saxelin, M., Pessi, T., and Salminen, S. Fecal recovery following oral administration of *Lactobacillus* GG (ATCC 53103) in gelatin capsules to healthy volunteers. *Int. J. Food Microbiol.* **25:** 199–203, 1995.
8. Wesney, E., and Tannock, G. W. Association of rat, pig and fowl biotypes of lactobacilli with the stomach of gnotobiotic mice. *Microb. Ecol.* **5:** 35–42, 1979.

2.3 MAINTENANCE OF PROBIOTIC MICROORGANISMS[1,2]

Organism	Period of Storage	Culture Medium/ Method	Growth Phase	Storage Temperature (°C)	Cryo-protectant	Frequency of Transfer
Lactobacilli	Short term	MRS agar stab	Early exponential	4–7°C	MRS agar	Weekly
	Short term	MRS agar spread	Early exponential	–20°C	MRS	Several months
	Long term	MRS	Late exponential	Lyophilized, 5–8°C	Skim milk or horse serum + 7.5% w/v glucose	10–20 years
Bifidobacteria	Short term	MRS 0.5% agar stab	Early exponential	3–4°C	MRS 0.5% agar stab	2 weeks
Bifidobacterium thermophilum	Short term	MRS agar slant in 10% CO_2 in air	Early exponential	3–4°C	MRS agar slant	2 weeks
Streptococci	Short term	Appropriate medium slope/stab	Early exponential	4°C	Culture medium	Weekly
		Litmus milk + 10% chalk + 0.3% yeast extract + 10% glucose	Early exponential	10°C	Litmus milk	Weekly
	Long term	Appropriate medium	Late exponential	–80°C deep frozen or in liquid N_2	1% tryptone, 0.5% yeast extract, 0.1% glucose, 0.1% cystein HCl, 2% bovine serum	Several months
	Long term	Appropriate medium	Late exponential	Lyophilized, 5–8°C	Skim milk or serum	10–20 years
Enterococci	Short term	Litmus milk + 1% chalk	Early exponential	4°C	Litmus milk + 1% chalk	3 months

(continued)

Organism	Period of Storage	Culture Medium/ Method	Growth Phase	Storage Temperature (°C)	Cryo-protectant	Frequency of Transfer
Enterococci (*cont.*)	Long term	Appropriate medium	Late exponential	Lyophilized –20°C	Nutrient broth + inactivated horse serum + glucose	2 years

[a]After Butler, J. P. (ed.), *Bergey's Manual of Systematic Bacteriology,* Vol. 2. Ed. Williams & Wilkins, Baltimore. 1986.

[b]Balows, A., Trüper, H., Devorkin, M., Harder, W., and Schleifer, K.-H. (eds.), *The Prokaryotes. A Handbook on the Biology of Bacteria: Ecophysiology, Isolation, Identification, Applications,* Vol. 2. Springer-Verlag, New York, 1992.

3

GENETIC MODIFICATION OF PROBIOTIC MICROORGANISMS

3.1 MUTANTS OBTAINED FROM LACTIC ACID BACTERIA BY MUTAGENESIS WITH NITROSOGUANIDINE OR ETHANE SULFONATE ETHYL ESTER

Species	Concentration of Agents	Character Acquired by NTG[a] Treatment	Ref.
NTG (mM)			
Lactobacillus plantarum	500	Defective producing ability of CO_2 from malate	2
	100–500	Rifampicin resistance	13
		Streptomycin resistance	14
	100	Ability to synthesize amino acid	10
L. casei	400	Increased ability to degrade protein	15
	50–500	Increased ability to degrade protein	12
	100–500	Increased ability to produce flavor and acid from soy-milk	9
	100	Ability to synthesize amino acid	10
L. helveticus	100	Ability to synthesize amino acid	10
L. acidophilus	100	Ability to synthesize amino acid	10
Streptococcus thermophilus	50	Increased acid productivity	6
	100–500	Ability of fermenting galactose	1
	500	Defective ability to transport lactose or sucrose	16
Enterococcus faecalis	300	Defective H^+-ATPase	17
	200	Defective Na-ATPase	7

(*continued*)

Species	Concentration of Agents	Character Acquired by NTG[a] Treatment	Ref.
Enterococcus faecalis (*cont.*)	50	Defective sugar-fermenting ability	5
	100	Ability to synthesize amino acid	3
Pediococcus acidilactici	100	Ability to synthesize amino acid	3
Leuconostoc oenos	500	Defective malolactic enzyme	8
Leuconostoc cremoris	100–600	Increased ability to produce acetoin and diacetyl	11
EMS[a] (%)			
L. casei	6	Defective peptidase activity	4
	2	Ability to synthesize amino acid	10
L. plantarum	2	Ability to synthesize amino acid	10
L. helveticus	2	Ability to synthesize amino acid	10
L. acidophilus	2	Ability to synthesize amino acid	10
E. faecalis	1.5	Defective sugar metabolism	5
E. faecium	2	Ability to synthesize amino acid	3
P. acidilactici	2	Ability to synthesize amino acid	3

[a]Abbreviations: NTG, *N*-methyl-*N′*-nitro-*N*-nitrosoguanidine; EMS, ethane sulfonate ethyl ester.

REFERENCES

1. Benateya, A., Bracquart, P., and Linden, G. Galactose- fermenting mutants of *Streptococcus thermophilus*. *Can. J. Microbiol.* **37:** 136–140, 1991.
2. Daeschel, M. A., McFeeters, R. F., Fleming, H. P., Klaenhammer, T. R., and Sanozky, R. B. Mutation and selection of *Lactobacillus plantarum* strains that do not produce carbon dioxide from malate. *Appl. Environ. Microbiol., 47:* 419–420, 1984.
3. Deguchi, Y., and Morishita, T. Nutritional requirements in multiple auxotrophic lactic acid bacteria: Genetic lesions affecting amino acid biosynthetic pathways in *Lactococcus lactis, Enterococcus faecium,* and *Pediococcus acidilactici. Biosci. Biotechnol. Biochem.* **56:** 913–918, 1992.
4. El Abboud, M., El Soda, M., Johnson, M., Olson, N. F., Simard, R. E., and Pandian, S. Peptidase deficient mutants—a new tool for the study of cheese ripening. *Milchwissenschaft* **47:** 625–628, 1992.
5. Feary, T. W., and Mayo, J. A. Detection of streptococcal mutants presumed to be defective in sugar catabolism. *Appl. Environ. Microbiol.* **47:** 1348–1351, 1984.
6. Hosono, A., Zhang, X., and Otani, H. Isolation of spontaneous induced-mutants of *Streptococcus thermophilus* having a high ability to produce acid. *Jpn. J. Dairy Food Sci.* **38:** A-9–A-13, 1989.
7. Kakinuma, Y., and Igarshi, K. Mutants of *Streptococcus faecalis* sensitive to alkaline pH lack Na^+-ATPase. *J. Bacteriol.* **172:** 1732–1735, 1990.
8. Lautensach, A., and Subden, R. E. Isolation and characterization of *Leuconostoc oenos* mutants with defective malo-lactic enzyme. *Microbios* **52:** 115–127, 1987.
9. Miyamoto, T., Reddy, N. S., and Nakae, T. Induction of mutation in *Lactobacillus casei* subsp. *alactosus* by nitrosoguanidine. *Agric. Biol. Chem.* **47:** 2755–2759, 1983.

10. Morishita, T., Deguchi, Y., Yajima, M., Sakurai, T., and Yura, T. Multiple nutritional requirements of lactobacilli: Genetic lesions affecting amino acid biosynthetic pathways. *J. Bacteriol.* **148:** 64–71, 1981.

11. Oberman, H., Libudzisz, Z., and Piatkiewicz, A. Mutation for *Leuconostoc cremoris* strains by nitrosoguanidine and ulraviolet irradiation. *Milchwisseschaft* **38:** 731–735, 1983.

12. Piatkiewicz, A., and Oberman, H. Influence of UV-rays and nitrosoguanidine on the proteolytic activity of *Lactobacillus casei. Acta Aliment. Pol.* **4:** 217–228, 1978.

13. Rodriguez-Quinones, F., Megias, M., Palomares, A. J., and Ruiz-Berranquero, F. Nitrosoguanidine mutagenesis in *Lactobacillus plantarum. Microbios Lett.* **20:** 75–80, 1982.

14. Rodriguez-Quinones, F., Palomares, A. J., Megias, M., and Ruiz-Berranquero, F. The influence of several variables for nitrosoguanidine mutagenesis in *Lactobacillus plantarum. Curr. Microbiol.* **10:** 137–140, 1984.

15. Singh, J., and Ranganathan, B. Activation of proteolytic activity of *Lactobacillus casei* by nitrosoguanidine. *Folia Microbiol.* **23:** 82–83, 1978.

16. Somkuti, G. A., and Steinberg, D. H. (-Fructofransidase activity in disaccharide transport mutants of *Streptococcus thermophilus. Biotechnol. Lett.* **13:** 809–814, 1991.

17. Suzuki, T., Unemoto, T., and Kobayashi, H. Novel streptococcal mutants defective in the regulation of H^+-ATPase biosynthesis and in F O complex. *J. Biol. Chem.* **263:** 11840–11843, 1988.

3.2 PLASMIDS

Strain	Plasmid	Size (kb)	Identified Gene	Phenotype[a]	Ref.
Bifidobacterium breve					
ATCC 15698	pNBb1	5.6			12
B. longum					
B2577	pMB1	1.9		Cryptic	75, 97
BK51	pTB6	3.6			74
KJ	pKJ36, pKJ50	3.6, 5.0			88
E. faecalis					
226 NWC	pEF226	5.2		Bac^{+}, Cnj	99
39-5	pPD1	59	*ipd, traA, traC, traB*	Bac^{+}, Phe	85, 128
CG180	pAM180	41		Em^{r}, Tc^{r}	123
D434	pIP1141	57.7		Cnj	45
DS16	pAD1	60	*iad, traA, traB, cylLL, cylLS, cylM, cylB*	Hly^{+}, Bac^{+}, Phe, Cnj, Cyt^{+}	23, 35, 63, 125
DS5	pAMγ1	53		Hly^{+}, Bac^{+}	63
	pAMβ1	26.5		Em^{r}	62, 110
	pAMα1	9.6		Tc^{r}	89
HH22	pBEM10	70		Bla^{+}, Gm^{r}, Phe	44, 81

(continued)

Strain	Plasmid	Size (kb)	Identified Gene	Phenotype[a]	Ref.
E. faecalis (*cont.*)					
	pAM323	66		Em^r, Phe	81
	pAM324	53		Phe	81
JH1	pJH1	80.7	*tetL*	Em^r, Tc^r, Km^r, Cnj	90, 116
	pJH2	58		Hly$^+$, Bac$^+$, Cnj	116
RC73	pAM373	36		Phe, Cnj	22, 83
S-48	pMB2	68		PA$^+$, Phe, Cnj	71, 95
	pMB1	90		Bac$^+$, Cnj, Phe	72, 95
	pMB3, pMB4	5.1, 2.5			95
SF-7	pCF10	58	*prgA, prgB, prgR, prgS, prgT, prgX, prgW, prgZ, prgY, prgQ*	Tc^r, Phe, Cnj	20, 21, 42, 84
YI717	pYI17	57.5	*bacA, bacB*	Bac$^+$, Phe	115
E. faecium					
226	pMBB1	2.9		Cryptic	127
228	pHKK100	55		Hly$^+$, Vm^r, Phe, Cnj	38
BM4102	pIP810	24.9		Em^r, Tm^r, Cnj	17
BM4147	pIP816	34	*vanA, vanY, vanH, vanX, vanS, vanR*	Vm^r	5, 14, 126
BM4165	pIP819	34	*ermAM*	Em^r Vm^r, Cnj	64
	pIP820	50	*tetM*	Pc^r, Tc^r	64
BM4178	pIP821	40		Vm^r, Cnj	64
	pIP822	4.5	*ermAM, tetM*	Pc^r	64
SC4	pKQ1	58.7	*tetM*	Tc^r, Em^r	30
L. acidophilus					
168 S	p1, p2, p3, p4, p5	1.6, 4, 4.2, 12, 60			25
ADH	pTRK15	26.5		Cryptic	70
LTF 42	pLA421	57.0	*lacG*	Lac$^+$	51
M46	pCV461	14	*acdB*	Bac$^+$	65, 118
RNL5	pLNL5	5		Cryptic	86
TK8912	pLA101	63		Gal$^+$	56
	pLA102	26.8	*lacG*	Lac$^+$	57
	pLA103	14	*acdT*	Bac$^+$	4, 54, 55
	pLA105	3.2		Cryptic	58
	pLA106	2.8			100
TK9201	pLA9201	45	*acdA*	Bac$^+$	53
strain 1899		4.2, 4.1			61
strain 301		4.2			61
strain NCFM		4.2			61

(*continued*)

Strain	Plasmid	Size (kb)	Identified Gene	Phenotype[a]	Ref.
strain R		3.5, 2.4, 2.1			61
L. bulgaricus					
10	pLB10	2.7			19
L. casei					
61BG	pLZ61	42		Lac^+	67
64H	pLZ64	35	*lacT, lacE, lacG, lacF*	Lac^+	3, 91
ATCC 11578	pLZ19A	65		Lac^+	67
	pLZ19B, pLZ19C, pLZ19D, pLZ19E, pLZ19F	6.8, 6.7 5.8 4.9, 4.8		Cryptic	67
ATCC 334	pLZ14A	28		Lac^+	67
	pLZ14B, pLZ14C	11, 9.3		Cryptic	67
ATCC 393	pLZ15	27		Lac^+	67
ATCC 4646	pLZ18A	55		Lac^+	67
	pLZ18B, pLZ18C	11.7, 6.4		Cryptic	67
C257	pLY101	68.2	*lacG*	Lac^+	102, 103
CG11		30		Muc^+	60
NCDO151	pNCDO151	15		Cryptic	130
O12		39.5, 25.5, 6.8			87
U14		39.5, 25.5			87
L. casei ssp. *casei*					
NCIB4114		6.9		Muc^+	121
L. curvatus					
IFPL 105		46		Bac^+	18
LTH1174		60	*curA*	Bac^+	114
LTH683	pLC2	2.5		Cryptic	59, 122
L. delbruckii ssp. *lactis*					
WS97	pWS97	60		Cryptic	130
L. fermentum					
LEM89	pLEM3	5.7	*erm*	Em^r	31
LF601	pLY2	15.6		Tc^r	47
	pLY4	57.8		Em^r	47
L. gasseri					
CNRZ222		150		Linear plasmid	98
IP102991		50		Linear plasmid	98
L. helveticus					
481	pMJ1008	12.6			48
ATCC 15009	pLH1, pLH2, pLH3	22.0, 6.0, 3.5		Cryptic	32, 94
CNRZ 1904		34		R/M	28
CP53	pCP53	11.5			129

(continued)

Strain	Plasmid	Size (kb)	Identified Gene	Phenotype[a]	Ref.
HLM1		5.5		Prt⁺	78
		31.5, 11.3, 7.6			78
LBL4	pLH4	2.6		Cryptic	94
L. helveticus ssp. *jugurti*					
S36.2	pLHJ1	13.2		Prt⁺	29
SBT2161	pLJ1	3.3		Cryptic	111
L. hilgardii					
67	pLAB1000	3.3		Cryptic	49, 50
	pLAB2000	9.1			49
L. pentosus					
MD353	p353-1	1.7			92
	p353-2	2.4		Cryptic	66, 93
L. plantarum					
A112	pA1	2.8		Cryptic	124
ATCC 8014	p8014-2	1.9		Cryptic	66
C3.8			*lacZ*	Lac⁺	76
C30I	pC30I1	2.1		Cryptic	105
CCM 1904	pLP1	2.1		Cryptic	11, 13
IFO 3070		11		Cit⁺	82
K40		55.5, 5.5, 5.3, 5.0, 3.1			87
LTF154	pLP1542	15		Bac⁺	52
	pLP1541, pLP1543	4.4, 63			52
NC1		7.1		Cryptic	104
NC4		44.1, 10.4, 9.5, 2.2		Cryptic	104
NC5		63, 20, 16.9, 12.3		Cryptic	104
NCDO1088	pLB4	3.5		Cryptic	9
W28		47.5			87
W37		55.5			87
caTC2R	pCaT	8.5	*cat*	Cmʳ	1
L. reuteri					
100-63	pGT633	9.8	*ermGT*	Emʳ	113
1044	pLU631	10.2	*erm*	Emʳ	6
1048		8.5	*erm*	Emʳ	6
1063	pLU631	10.2	*erm*	Emʳ	6
1068	pLU631	10.2	*erm*	Emʳ	6
G4	pTC82	7.0	*cat-TC*	Cmʳ	68
L. sake					
L45	pCIM1	50	*lasA, lasM, lasT, lasP*	Bac⁺	79,106, 107
	pCM2	34			106
Lb706	pLSA60	60	*sapA, saiA, sapK, sapR, sapT, sapE*	Bac⁺	7, 8

(*continued*)

Strain	Plasmid	Size (kb)	Identified Gene	Phenotype[a]	Ref.
Lactobacillus					
sp. strain 100-33	pLAR33	18	*erm*	Em^r	96
ssp. DB27		53.1		Mal^+	69
ssp. DB28		53.1		Mal^+	69
ssp. DB29	pML291	75.5		Mal^+	69
ssp. DB31		53.1		Mal^+	69
strain 25	pSE601	2.8		Cryptic	2
strain 65		11.5		Muc^+	2
strain 80		61.4		Sor^+	2
Leuconostoc gelium					
UAL187	pLG7.6	12	*lcaE, lcaC, lcaD*	Bac^+	41, 117
	pLG9.2	14.5			40, 117
Leuconostoc lactis					
NZ6009	pNZ63	23	*lacL, lacM*	Lac^+	27
533	pCI411	2.7		Cryptic	24
NZ6070	pCT71		*citP*	Cit^+	120
Leuconostoc mesenteroides					
Y105	pHY30	35	*mesY, mesI, mesC, mesD, mesE*	*Bac+*	34
	pHY15, pHY3, pHY2	15, 3.5, 2.5			34
Leuconostoc oenos					
8413	pLo13	4		Cryptic	33
ATCC 23277	pLoA2, ppLoA1	40, 4.3			33
CECT 4028	p4028	4.4		Cryptic	131
bOg32	pOg32	2.5		Cryptic	15
Leuconostoc paramesenteroides					
NZ6009	pNZ63	26	*lacS*	Lac^+	119
	pNZ64, pNZ62, pNZ61	35, 18.5, 1.8			26
P. acidilactici					
HM3020	pVM20	46	*ermAM*	Em^r	112
	pVM21	4			112
LB42-923	pSMB74	8.9	*papA, papB, papC, papD*	Bac^+	16, 80
PAC1.0	pSRQ11	9.4	*pedA, pedB, papC, papD*	Bac^+	73
PAC1.0	pSQ10	36		Suc^+	37
SJ-1		7		Bac^+	101
UL5	pMJ5	12.9		Bac^+	46
P. pentosaceus					
NCDO 990		37, 10.5			10
P. pentosus					
PPE1.0	pSRQ1	47		Suc^+, Raf^+, Mel^+	36

(*continued*)

Strain	Plasmid	Size (kb)	Gene	Phenotype[a]	Ref.
PPE2.0	pSRQ12	44		Suc$^+$, Raf$^+$, Mel$^+$	36
	pSRQ13, pSRQ14, pSRQ15, pSRQ16	30, 13, 9.5 3.6			36
PPE5.0	pSRQ17	36		Suc$^+$, Raf$^+$, Mel$^+$	36
	pSRQ18, pSRQ19, pSRQ20	11, 9.8, 1.9			36
S. thermophilus					
3641	pHM1	2.2		Cryptic	43
AO33	pA33	6.9		Mor$^+$	77
CR2ST	pHM4	3.2		Cryptic	43
No. 29	pST1	2.8		Cryptic	39
R2ST	pHM3	3.5		Cryptic	43
ST108	pER8	2.2			108, 109
ST113	pER13	4.2			108
ST116	pER16	4.5			108
ST134	pER341, pER342	2.7, 9.5			108
ST136	pER36	3.7			108
ST2	pHM2	3.2		Cryptic	43
STA	pHM5	2.2		Cryptic	43
Tetragenococcus halophilia (Pediococcus halophilus)					
ATCC 33315	pUCL287	8.7		Cryptic	10

[a]Abbreviations

Bac$^+$, bacteriocin production
Bla$^+$, β-lactamase production
Cit$^+$, citrate fermentation
Cm^r, chloramphenicol resistance
Cnj, conjugative transfer
Cyt$^+$, cytolysin production
Em^r, erythromycin resistance
Gal$^+$, galactose fermentation
Gm^r, gentamicin resistance
Hly$^+$, hemolysin production
Km^r, kanamycin resistance
Lac$^+$, lactose fermentation
Mal$^+$, maltose fermentation
Mel$^+$, melibiose fermentation
Mor$^+$, change in morphology
Muc, mucoid expression
PA$^+$, peptide antibiotic production
Pc^r, penicillin resistance
Phe, pheromone response
Prt$^+$, proteolytically active
Raf, raffinose fermentation
R/M, restriction–modification system
Sor$^+$, sorbitol fermentation
Suc$^+$, sucrose fermentation
Tc^r, tetracycline resistance
Tm^r, tobramycin resistance
Vm^r, vancomycin resistance

REFERENCES

1. Ahn, C., Collins-Thompson, D., Duncan, C., and Stiles, M. E. Mobilization and location of the genetic determinant of chloramphenicol resistance from *Lactobacillus plantarum* caTC2R. *Plasmid* **27:** 169–176, 1992.
2. Ahrne, S., Molin, G., and Stahl, S. Plasmids in *Lactobacillus* strains isolated from meat and meat products. *Syst. Appl. Microbiol.* **11:** 320–325, 1989.

3. Alpert, C. A., and Siebers, U. The *lac* operon of *Lactobacillus casei* contains *lacT*, a gene coding for a protein of the BglG family of transcriptional antiterminator. *J. Bacteriol.* **179:** 1555–1562, 1997.
4. Argnani, A., Leer, R. J., van Luijk, N., and Pouwels, P. H. A convenient and reproducible method to genetically transform bacteria of the genus *Bifidobacterium. Microbiology* **142:** 109–114, 1996.
5. Arthur, M., Molinas, C., and Courvalin, P. The VanS-VanR two-component regulatory system controls synthesis of depsipeptide peptidoglycan precursors in *Enterococcus faecium* BM4147. *J. Bacteriol.* **174:** 2582–2591, 1992.
6. Axelsson, L. T., Ahrne, S. E., Andersson, M. C., and Stahl, S. R. Identification and cloning of a plasmid-encoded erythromycin resistance determinant from *Lactobacillus reuteri. Plasmid* **20:** 171–174, 1988.
7. Axelsson, L., and Holck, A. The genes involved in the production of and immunity to sakacin A, a bacteriocin from *Lactobacillus sake* Lb706. *J. Bacteriol.* **177:** 2125–2137, 1995.
8. Axelsson, L., Holck, A., Birkeland, S. E., Aukrust, T., and Blom, H. Cloning and nucleotide sequence of a gene from *Lactobacillus sake* Lb706 necessary for sakacin A production and immunity. *Appl. Environ. Microbiol.* **59:** 2868–2875, 1993.
9. Bates, E. E., and Gilbert, H. J. Characterization of a cryptic plasmid from *Lactobacillus plantarum. Gene* **85:** 253–258, 1989.
10. Benachour, A., Frere, J., and Novel, G. pUCL287 plasmid from *Tetragenococcus halophila (Pediococcus halophilus)* ATCC 33315 represents a new theta-type replicon family of lactic acid bacteria. *FEMS Microbiol. Lett.* **128:** 167–176, 1995.
11. Bouia, A., Bringel, F., Frey, L., Kammerer, B., Belarbi, A., Guyonvarch, A., and Hubert, J. C. Structural organization of pLP1, a cryptic plasmid from *Lactobacillus plantarum* CCM 1904. *Plasmid* **22:** 185–192, 1989.
12. Bourget, N., Simonet, J. M., and Decaris, B. Analysis of the genome of the five *Bifidobacterium breve* strains: Plasmid content, pulsed-field gel electrophoresis genome size estimation and rrn loci number. *FEMS Microbiol. Lett.* **110:** 11–20, 1993.
13. Bringel, F., Frey, L., and Hubert, J. C. Characterization, cloning, curing, and distribution in lactic acid bacteria of pLP1, a plasmid from *Lactobacillus plantarum* CCM 1904 and its use in shuttle vector construction. *Plasmid* **22:** 193–202, 1989.
14. Brisson-Noel, A., Dutka-Malen, S., Molinas, C., Leclercq, R., and Courvalin, P. Cloning and heterospecific expression of the resistance determinant vanA encoding high-level resistance to glycopeptides in *Enterococcus faecium* BM4147. *Antimicrob. Agents Chemother.* **34:** 924–927, 1990.
15. Brito, L., Vieira, G., Santos, M. A., and Paveia, H. Nucleotide sequence analysis of pOg32, a cryptic plasmid from *Leuconostoc oenos. Plasmid* **36:** 49–54, 1996.
16. Bukhtiyarova, M., Yang, R., and Ray, B. Analysis of the pediocin AcH gene cluster from plasmid pSMB74 and its expression in a pediocin-negative *Pediococcus acidilactici* strain. *Appl. Environ. Microbiol.* **60:** 3405–3408, 1994.
17. Carlier, C., and Courvalin, P. Emergence of 4′,4″-aminoglycoside nucleotidyltransferase in enterococci. *Antimicrob. Agents Chemother.* **34:** 1565–1569, 1990.
18. Casla, D., Requena, T., and Gomez, R. Antimicrobial activity of lactic acid bacteria isolated from goat's milk and artisanal cheeses: Characteristics of a bacteriocin produced by *Lactobacillus curvatus* IFPL 105. *J. Appl. Bacteriol.* **81:** 35–41, 1996.

19. Chagnaud, P., Chan, K. C., Duran, R., Naouri, P., Arnaud, A., and Galzy, P. Construction of a new shuttle vector for *Lactobacillus. Can. J. Microbiol.* **38:** 69–74, 1992.

20. Christie, P. J., and Dunny, G. M. Identification of regions of the *Streptococcus faecalis* plasmid pCF-10 that encode antibiotic resistance and pheromone response functions. *Plasmid* **15:** 230–241, 1986.

21. Chung, J. W., Bensing, B. A., and Dunny, G. M. Genetic analysis of a region of the *Enterococcus faecalis* plasmid pCF10 involved in positive regulation of conjugative transfer functions. *J. Bacteriol.* **177:** 2107–2017, 1995.

22. Clewell, D. B., An, F. Y., White, B. A., and Gawron-Burke, C. *Streptococcus faecalis* sex pheromone (cAM373) also produced by *Staphylococcus aureus* and identification of a conjugative transposon (Tn918). *J. Bacteriol.* **162:** 1212–1220, 1985.

23. Clewell, D. B., Pontius, L. T., An, F. Y., Ike, Y., Suzuki, A., and Nakayama, J. Nucleotide sequence of the sex pheromone inhibitor (iAD1) determinant of *Enterococcus faecalis* conjugative plasmid pAD1. *Plasmid* **24:** 156–161, 1990.

24. Coffey, A., Harrington, A., Kearney, K., Daly, C., and Fitzgerald, G. Nucleotide sequence and structural organization of the small, broad-host-range plasmid pCI411 from *Leuconostoc lactis* 533. *Microbiology* **140:** 2263–2269, 1994.

25. Damiani, G., Romagnoli, S., Ferretti, L., Morelli, L., Bottazzi, V., and Sgaramella, V. Sequence and functional analysis of a divergent promoter from a cryptic plasmid of *Lactobacillus acidophilus* 168 S. *Plasmid* **17:** 69–72, 1987.

26. David, S., Simons, G., and de Vos, W. M. Plasmid transformation by electroporation of *Leuconostoc paramesenteroides* and its use in molecular cloning. *Appl. Environ. Microbiol.* **55:** 1483–1489, 1989.

27. David, S., Stevens, H., van Riel, M., Simons, G., and de Vos, W. M. *Leuconostoc lactis* beta-galactosidase is encoded by two overlapping genes. *J. Bacteriol.* **174:** 4475–4481, 1992.

28. de Los Reyes-Gavilan, C. G., Limsowtin, G. K. Y., Séchaud, L., Veaux, M., and Accolas, J. P. Evidence for a plasmid-linked restriction-modification system in *Lactobacillus helveticus. Appl. Environ. Microbiol.* **56:** 3412–3419, 1990.

29. de Rossi, E., Brigidi, P., Riccardi, G., Milano, A., and Matteuzzi, D. Preliminary studies on the correlation between the plasmid pHJ1 and its proteolytic activity in *Lactobacillus helveticus* S 36.2. Physical mapping and molecular cloning of the plasmid in Escherichia coli. *Microbiologica (Bologna)* **12:** 273–276, 1989.

30. Fletcher, H. M., and Daneo-Moore, L. A truncated Tn916-like element in a clinical isolate of *Enterococcus faecium. Plasmid* **27:** 155–160, 1992.

31. Fons, M., Hege, T., Ladire, M., Raibaud, P., Ducluzeau, R., and Maguin, E. Isolation and characterization of a plasmid from *Lactobacillus fermentum* conferring erythromycin resistance. *Plasmid* **37:** 199–203, 1997.

32. Fortina, M. G., Parini, C., Rossi, P., and Manachini, P.L. Mapping of three plasmids from *Lactobacillus helveticus* ATCC 15009. *Lett. Appl. Microbiol.* **17:** 303–306, 1993.

33. Fremaux, C., Aigle, M., and Lonvaud-Funel, A. Sequence analysis of *Leuconostoc oenos* DNA: Organization of pLo13, a cryptic plasmid. *Plasmid* **30:** 212–223, 1993.

34. Fremaux, C., Hechard, Y., and Cenatiempo, Y. Mesentericin Y105 gene clusters in *Leuconostoc mesenteroides* Y105. *Microbiology* **141:** 1637–1645, 1995.

35. Gilmore, M. S., Segarra, R. A., Booth, M. C., Bogie, C. P., Hall, L. R., and Clewell, D. B. Genetic structure of the *Enterococcus faecalis* plasmid pAD1-encoded cytolytic tox-

in system and its relationship to lantibiotic determinants. *J. Bacteriol.* **176:** 7335–7344, 1994.

36. Gonzalez, C. F., and Kunka, B. S. Evidence for plasmid linkage of raffinose utilization and associated a-galactosidase and sucrose hydrolase activity in *Pediococcus pentosus. Appl. Environ. Microbiol.* **51:** 105–109, 1986.
37. Gonzalez, C. F., and Kunka, B. S. Plasmid-associated bacteriocin production and sucrose fermentation in *Pediococcus acidilactici. Appl. Environ. Microbiol.* **53:** 2534–2538, 1987.
38. Handwerger, S., Pucci, M. J., and Kolokathis, A. Vancomycin resistance is encoded on a pheromone response plasmid in *Enterococcus faecium* 228. *Antimicrob. Agents Chemother.* **34:** 358–360, 1990.
39. Hashiba, H., Takiguchi, R., Joho, K., Aoyama, K., and Hirota, T. Identification of the replication region of *Streptococcus thermophilus* No. 29 plasmid pST1. *Biosci. Biotechnol. Biochem.* **57:** 1646–1649, 1993.
40. Hastings, J. W., and Stiles, M. E. Antibiosis of *Leuconostoc gelidum* isolated from meat. *J. Appl. Bacteriol.* **70:** 127–134, 1991.
41. Hastings, J. W., Sailer, M., Johnson, K., Roy, K. L., Vederas, J. C., and Stiles, M. E. Characterization of leucocin A-UAL 187 and cloning of the bacteriocin gene from *Leuconostoc gelidum. J. Bacteriol.* **173:** 7491–7500, 1991.
42. Hedberg, P. J., Leonard, B. A., Ruhfel, R. E., and Dunny, G. M. Identification and characterization of the genes of *Enterococcus faecalis* plasmid pCF10 involved in replication and in negative control of pheromone-inducible conjugation. *Plasmid* **35:** 46–57, 1996.
43. Herman, R. E., and McKay, L. L. Isolation and partial characterization of plasmid DNA from *Streptococcus thermophilus. Appl. Environ. Microbiol.* **50:** 1103–1106, 1985.
44. Hodel-Christian, S. L., and Murray, B. E. Mobilization of the gentamicin resistance gene in *Enterococcus faecalis. Antimicrob. Agents Chemother.* **34:** 1278–1280, 1990.
45. Horaud, T., Delbos, F., and de Cespedes, G. Tn3702, a conjugative transposon in *Enterococcus faecalis. FEMS Microbiol. Lett.* **72:** 189–194, 1990.
46. Huang, J., Lacroix, C., Daba, H., and Simard, R. E. Pediocin 5 production and plasmid stability during continuous free and immobilized cell cultures of *Pediococcus acidilactici* UL5. *J. Appl. Bacteriol.* **80:** 635–644, 1996.
47. Isiwa, H., and Iwata, S. Drug resistance plasmids in *Lactobacillus fermentum. J. Gen. Appl. Microbiol.* **26:** 71–74, 1980.
48. Joerger, M. C., and Klaenhammer, T. R. Characterization and purification of helveticin J and evidence for a chromosomally determined bacteriocin produced by *Lactobacillus helveticus* 481. *J. Bacteriol.* **167:** 439–446, 1986.
49. Josson, K., Scheirlinck, T., Michiels, F., Platteeuw, C., Stanssens, P., Joos, H., Dhaese, P., Zabeau, M., and Mahillon, J. Characterization of a gram-positive broad-host-range plasmid isolated from *Lactobacillus hilgardii. Plasmid* **21:** 9–20, 1989.
50. Josson, K., Soetaert, P., Michiels, F., Joos, H., and Mahillon, J. *Lactobacillus hilgardii* plasmid pLAB1000 consists of two functional cassettes commonly found in other gram-positive organisms. *J. Bacteriol.* **172:** 3089–3099, 1990.
51. Kanatani, K., and Oshimura, M. Isolation and structural analysis of the phospho-galactosidase gene from *Lactobacillus acidophilus. J. Ferment. Bioeng.* **78:** 123–129, 1994.

52. Kanatani, K., and Oshimura, M. Plasmid-associated bacteriocin production by a *Lactobacillus plantarum* strain. *Biosci. Biotechnol. Biochem.* **58:** 2084–2086, 1994.
53. Kanatani, K., Oshimura, M., and Sano, K. Isolation and characterization of acidocin A and cloning of the bacteriocin gene from *Lactobacillus acidophilus. Appl. Environ. Microbiol.* **61:** 1061–1067, 1995.
54. Kanatani, K., Tahara, T., Oshimura, M., Sano, K., and Umezawa, C. Identification of the replication region of *Lactobacillus acidophilus* plasmid pLA103. *FEMS Microbiol. Lett.* **133:** 127–130, 1995.
55. Kanatani, K., Tahata, T., Yoshida, K., Miura, H., Sakamoto, M., and Osimura, M. Plasmid-associated bacteriocin production by and immunity of *Lactobacillus acidophilus* TK8912. *Biosci. Biotechnol. Biochem.* **56:** 648–651, 1992.
56. Kanatani, K., Tahara, T., Yoshida, K., Miura, H., Sakamoto, M., and Oshimura, M. Plasmid-linked galactose utilization by *Lactobacillus acidophilus* TK8912. *Biosci. Biotechnol. Biochem.* **56:** 826–827, 1992.
57. Kanatani, K., Yoshida, K., Tahara, T., Miura, H., Sakamoto, M., and Oshimura, M. Isolation and characterization of plasmid DNA in *Lactobacillus acidophilus. Agric. Biol. Chem.* **55:** 2051–2056, 1991.
58. Kanatani, K., Yoshida, K., Tahara, T.,Yamada, K., Miura, H., Sakamoto, M., and Oshimura, M. Transformation of *Lactobacillus acidophilus* TK1892 by electroporation with pULA105E plasmid. *J. Ferment. Bioeng.* **74:** 358–362, 1992.
59. Klein, J. R., Ulrich, C., and Plapp, R. Characterization and sequence analysis of a small cryptic plasmid from *Lactobacillus curvatus* LTH683 and its use for construction of new *Lactobacillus* cloning vectors. *Plasmid* **30:** 14–29, 1993.
60. Kojic, M., Vujcic, M., Banina, A., Cocconcelli, P., Cerning, J., and Topisirovic, L. Analysis of exopolysaccharide production by *Lactobacillus casei* CG11 isolated from cheese. *Appl. Environ. Microbiol.* **58:** 4086–4088, 1992.
61. Kumar, R., Garg, S. K., Singh, D. T., Singh, S. P., and Mital, B. K. Evidence for the presence of plasmids in four therapeutically important strains of *Lactobacillus acidophilus. Lett. Appl. Microbiol.* **19:** 188–191, 1994.
62. LeBlanc, D. J., and Lee, L. N. Physical and genetic analyses of streptococcal plasmid pAMβ1 and cloning of its replication region. *J. Bacteriol.* **157:** 445–453, 1984.
63. LeBlanc, D. J., Lee, L. N., Clewell, D. B., and Behnke, D. Broad geographical distribution of a cytotoxin gene mediating beta-hemolysis and bacteriocin activity among *Streptococcus faecalis* strains. *Infect. Immun.* **40:** 1015–1022, 1983.
64. Leclercq, R., Derlot, E., Weber, M., Duval, J., and Courvalin, P. Transferable vancomycin and teicoplanin resistance in *Enterococcus faecium. Antimicrob. Agents Chemother.* **33:** 10–15, 1989.
65. Leer, R. J., van der Vossen, J. M., van Giezen, M., van Noort, J. M., and Pouwels, P. H. Genetic analysis of acidocin B, a novel bacteriocin produced by *Lactobacillus acidophilus. Microbiology* **141:** 1629–1635, 1995.
66. Leer, R. J., van Luijk, N., Posno, M., and Pouwels, P. H. Structural and functional analysis of two cryptic plasmids from *Lactobacillus pentosus* MD353 and *Lactobacillus plantarum* ATCC 8014. *Mol. Gen. Genet.* **234:** 265–274, 1992.
67. Lee-Wickner, L. J., and Chassy, B. M. Characterization and molecular cloning of cryptic plasmids isolated from *Lactobacillus casei. Appl. Environ. Microbiol.* **49:** 1154–1161, 1985.

68. Lin, C. F., Fung, Z. F., Wu, C. L., and Chung, T. C. Molecular characterization of a plasmid-borne (pTC82) chloramphenicol resistance determinant (cat-TC) from *Lactobacillus reuteri* G4. *Plasmid* **36:** 116–124, 1996.

69. Liu, M. L., Kondo, J. K., Barnes, M. B., and Bartholomew, D. T. Plasmid-linked maltose utilization in *Lactobacillus* ssp. *Biochimie* **70:** 351–355, 1988.

70. Luchansky, J. B., Muriana, P. M., and Klaenhammer, T. R. Application of electroporation for transfer of plasmid DNA to *Lactobacillus, Lactococcus, Leuconostoc, Listeria, Pediococcus, Bacillus, Staphylococcus, Enterococcus* and *Propionibacterium. Mol. Microbiol.* **2:** 637–646, 1988.

71. Martinez-Bueno, M., Galvez, A., Valdivia, E., and Maqueda, M. A transferable plasmid associated with AS-48 production in *Enterococcus faecalis. J. Bacteriol.* **172:** 2817–2818, 1990.

72. Martinez-Bueno, M., Valdivia, E., Galvez, A., and Maqueda, M. Transfer of a plasmid determining bacteriocin Bc-48 production and immunity, and response to sexual pheromones in *Enterococcus faecalis* S-48. *Plasmid* **28:** 61–69, 1992.

73. Marugg, J. D., Gonzalez, C. F., Kunka, B. S., Ledeboer, A. M., Pucci, M. J., Toonen, M. Y., Walker, S. A., Zoetmulder, L. C., and Vandenbergh, P. A. Cloning, expression, and nucleotide sequence of genes involved in production of pediocin PA-1, and bacteriocin from *Pediococcus acidilactici* PAC1.0. *Appl. Environ. Microbiol.* **58:** 2360–2367, 1992.

74. Matsumura, H., Takeuchi, A., and Kano, Y. Construction of *Escherichia coli-Bifidobacterium longum* shuttle vector transforming *B. longum* 105-A and 108-A. *Biosci. Biotechnol. Biochem.* **61:** 1211–1212, 1997.

75. Matteuzzi, D., Brigidi, P., Rossi, M., and Di, D. Characterization and molecular cloning of *Bifidobacterium longum* cryptic plasmid pMB1. *Lett. Appl. Microbiol.* **11:** 220–223, 1990.

76. Mayo, B., Gonzales, B., Arca, P., and Snarez, J. E. Cloning and expression of the plasmid encoded β-D-galactosidase gene from a *Lactobacillus plantarum* strain of dairy origin. *FEMS Microbiol. Lett.* **122:** 145–152, 1994.

77. Mercenier, A. Molecular genetics of *Streptococcus thermophilus. FEMS Microbiol. Rev.* **7:** 61–77, 1990.

78. Morelli, L., Vescovo, M., Cocconcelli, P. S., and Bottazzi, V. Fast and slow milk-coagulating variants of *Lactobacillus helveticus* HLM1. *Can. J. Microbiol.* **32:** 758–762, 1986.

79. Mortvedt, C. I., and Nes, I. F. Plasmid-associated bacterioin production by a *Lactobacillus sake* strain. *J. Gen. Microbiol.* **136:** 1601–1607, 1990.

80. Motlagh, A., Bukhtiyarova, M., and Ray, B. Complete nucleotide sequence of pSMB 74, a plasmid encoding the production of pediocin AcH in *Pediococcus acidilactici. Lett. Appl. Microbiol.* **18:** 305–312, 1994.

81. Murray, B. E., An, F. Y., and Clewell, D. B. Plasmids and pheromone response of the β-lactamase producer *Streptococcus (Enterococcus) faecalis* HH22. *Antimicrob. Agents Chemother.* **32:** 547–551, 1988.

82. Nakamura, S., Miyamoto, T., Izumimoto, M., and Kataoka, K. Isolation and characterization of citrate metabolism-deficient mutants in *Lactobacillus plantarum* IFO 3070. *Anim. Sci. Technol. Jpn.* **62:** 1142–1148, 1991.

83. Nakayama, J., Ono, Y., and Suzuki, A. Isolation and structure of the sex pheromone inhibitor, iAM373, of *Enterococcus faecalis. Biosci. Biotechnol. Biochem.* **59:** 1358–1359, 1995.

84. Nakayama, J., Ruhfel, R. E., Dunny, G. M., Isogai, A., and Suzuki, A. The prgQ gene of the *Enterococcus faecalis* tetracycline resistance plasmid pCF10 encodes a peptide inhibitor, iCF10. *J. Bacteriol.* **176:** 7405–7408, 1994.
85. Nakayama, J., Yoshida, K., Kobayashi, H., Isogai, A., Clewell, D. B., and Suzuki, A. Cloning and characterization of a region of *Enterococcus faecalis* plasmid pPD1 encoding pheromone inhibitor (ipd), pheromone sensitivity (traC), and pheromone shutdown (traB) genes. *J. Bacteriol.* **177:** 5567–5573, 1995.
86. Natori, Y., Kano, Y., and Imamoto, F. Characterization and promoter selectivity of *Lactobacillus acidophilus* RNA polymerase. *Biochimie* **70:** 1765–1774, 1988.
87. Olasupo, N. A., Olukoya, D. K., and Odunfa, S. A. Plasmid profiles of bacteriocin-producing *Lactobacillus* isolates from African fermented foods. *Folia Microbiol. (Prague)* **39:** 181–186, 1994.
88. Park, M. S., Lee, K. H., and Ji, G. E. Isolation and characterization of two plasmids from *Bifidobacterium longum. Lett. Appl. Microbiol.* **25:** 5–7, 1997.
89. Perkins, J. B., and Youngman, P. *Streptococcus* plasmid pAM alpha 1 is a composite of two separable replicons, one of which is closely related to *Bacillus* plasmid pBC16. *J. Bacteriol.* **155:** 607–615, 1983.
90. Platteeuw, C., Michiels, F., Joos, H., Seurinck, J., and de Vos, W. M. Characterization and heterologous expression of the tetL gene and identification of iso-ISS1 elements from *Enterococcus faecalis* plasmid pJH1. Gene **160:** 89–93, 1995.
91. Porter, E. V., and Chassy, B. M. Nucleotide sequence of the b-D-phosphogalactoside galactohydrolase gene of *Lactobacillus casei*: Comparison to analogous pbg genes of other gram-positive organisms. *Gene* **62:** 263–276, 1988.
92. Posno, M., Leer, R. J., van Luijk, N., van Giezen, M. J. F., Heuvelmans, P. T. H. M., Lokman, B. C., and Pouwels, P. H. Incompatibility of *Lactobacillus* vectors with replicons derived from small cryptic *Lactobacillus* plasmids and segregational instability of the introduced vectors. *Appl. Environ. Microbiol.* **57:** 1822–1828, 1991.
93. Pouwels, P. H., van Luijk, N., Leer, R. J., and Posno, M. Control of replication of the *Lactobacillus pentosus* plasmid p353-2: Evidence for a mechanism involving tanscriptional attenuation of the gene coding for the replication protein. *Mol. Gen. Genet.* **242:** 614–22, 1994.
94. Pridmore, D., Stefanova, T., and Mollet, B. Cryptic plasmids from *Lactobacillus helveticus* and their evolutionary relationship. *FEMS Microbiol. Lett.* **124:** 301–306, 1994.
95. Quirantes, R., Galvez, A., Valdivia, E., Martin, I., Martinez-Bueno, M., Mendez, E., and Maqueda, M. Purification of sex pheromones specific for pMB1 and pMB2 plasmids of *Enterococcus faecalis* S-48. *Can. J. Microbiol.* **41:** 629–632, 1995.
96. Rinckel, L. A., and Savage, D. C. Characterization of plasmids and plasmid-borne macrolide resistance from *Lactobacillus* sp. strain 100-33. *Plasmid* **23:** 119–125, 1990.
97. Rossi, M., Brigidi, P., Gonzalez-Vara-y-Rodriguez, A., and Matteuzzi, D. Characterization of the plasmid pMB1 from *Bifidobacterium longum* and its use for shuttle vector construction. *Res. Microbiol.* **147:** 133–143, 1996.
98. Roussel, Y., Colmin, C., Simonet, J. M., and Decaris, B. Strain characterization, genome size and plasmid content in the *Lactobacillus acidophilus* group (Hansen and Mocquot). *J. Appl. Bacteriol.* **74:** 549–556, 1993.
99. Salzano, G., Villani, F., Pepe, O., Sorrentino, E., Moschetti, G., and Coppola, S. Conju-

gal transfer of plasmid-borne bacteriocin production in *Enterococcus faecalis* 226 NWC. *FEMS Microbiol. Lett.* **99:** 1–6, 1992.

100. Sano, K., Otani, M., Okada, Y., Kawamura, R., Umesaki, M., Ohi, Y., Umezawa, C., and Kanatani, K. Identification of the replication region of the *Lactobacillus acidophilus* plasmid pLA106. *FEMS Microbiol. Lett.* **148:** 223–226, 1997.
101. Schved, F., Lalazar, A., Henis, Y., and Juven, B. J. Purification, partial characterization and plasmid-linkage of pediocin SJ-1, a bacteriocin produced by *Pediococcus acidilactici. J. Appl. Bacteriol.* **74:** 67–77, 1993.
102. Shimizu-Kadota, M. Properties of lactose plasmid pLY101 in *Lactobacillus casei. Appl. Environ. Microbiol.* **53:** 2987–2991, 1987.
103. Shimizu-Kadota, M. Cloning and expression of the phospho-beta-galactosidase genes on the lactose plasmid and the chromosome of *Lactobacillus casei* C257 in *Escherichia coli. Biochimie* **70:** 523–529, 1988.
104. Shrago, A. W., and Dobrogosz, W. J. Conjugal transfer of group B streptococcal plasmids and comobilization of *Escherichia coli-Streptococcus* shuttle plasmids to *Lactobacillus plantarum. Appl. Environ. Microbiol.* **54:** 824–826, 1988.
105. Skaugen, M. The complete nucleotide sequence of a small cryptic plasmid from *Lactobacillus plantarum. Plasmid* **22:** 175–179, 1989.
106. Skaugen, M., and Nes, I. F. Transposition in *Lactobacillus sake* and its abolition of lactocin S production by insertion of IS1163, a new member of the IS3 family. *Appl. Environ. Microbiol.* **60:** 2818–2825, 1994.
107. Skaugen, M., Abildgaard, C. I., and Nes, I. F. Organization and expression of a gene cluster involved in the biosynthesis of the lantibiotic lactocin S. *Mol. Gen. Genet.* **253:** 674–686, 1997.
108. Somkuti, G. A., and Steinberg, D. H. Genetic transformation of *Streptococcus thermophilus* by electroporation. *Biochimie* **70:** 579–585, 1988.
109. Somkuti, G. A., Solaiman, D. K., and Steinberg, D. H. Native promoter-plasmid vector system for heterologous cholesterol oxidase synthesis in *Streptococcus thermophilus. Plasmid* **33:** 7–14, 1995.
110. Swinfield, T. J., Oultram, J. D., Thompson, D. E., Brehm, J. K., and Minton, N. P. Physical characterisation of the replication region of the *Streptococcus faecalis* plasmid pAM beta 1. *Gene* **87:** 79–90, 1990.
111. Takiguchi, R., Hashiba, H., Aoyama, K., and Ishii, S. Complete nucleotide sequence and characterization of a cryptic plasmid from *Lactobacillus helveticus* subsp. *jugurti. Appl. Environ. Microbiol.* **55:** 1653–1655, 1989.
112. Tankovic, J., Leclercq, R., and Duval, J. Antimicrobial susceptibility of *Pediococcus* spp., and genetic basis of macrolide resistance in *Pediococcus acidilactici* HM3020. *Antimicrob. Agents Chemother.* **37:** 789–792, 1993.
113. Tannock, G. W., Luchansky, J. B., Miller, L., Connell, H., Thode-Andersen, S., Mercer, A. A., and Klaenhammer, T. R. Molecular characterization of a plasmid-borne (pGT633) erythromycin resistance determinant (ermGT) from *Lactobacillus reuteri* 100-63. *Plasmid* **31:** 60–71, 1994.
114. Tichaczek, P. S., Vogel, R. F., and Hammes, W. P. Cloning and sequencing of curA encoding curvacin A, the bacteriocin produced by *Lactobacillus curvatus* LTH1174. *Arch. Microbiol.* **160:** 279–283, 1993.
115. Tomita, H., Fujimoto, S., Tanimoto, K., and Ike, Y. Cloning and genetic organization of

the bacteriocin 31 determinant encoded on the *Enterococcus faecalis* pheromone-responsive conjugative plasmid pYI17. *J. Bacteriol.* **178:** 3585–3593, 1996.

116. Trieu-Cuot, P., and Courvalin, P. Nucleotide sequence of the *Streptococcus faecalis* plasmid gene encoding the 3'5''-aminoglycoside phosphotransferase type III. *Gene* **23:** 331–341, 1983.
117. van Belkum, M. J., and Stiles, M. E. Molecular characterization of genes involved in the production of the bacteriocin leucocin A from *Leuconostoc gelidum. Appl. Environ. Microbiol.* **61:** 3573–3579, 1995.
118. van der Vossen, J. M. B. M., van Herwijnen, M. H. M., Leer, R. J., ten Brink, B., Pouwels, P. H., and Huis I'nt Veld, J. H. J. Production of acidocin B, a bacteriocin of *Lactobacillus acidophilus* M46 is a plasmid-encoded trait: Plasmid curing, genetic marking by in vivo plasmid integration, and gene transfer. *FEMS Microbiol Lett.* **116:** 333–340, 1994.
119. Vaughan, E. E., David, S., and de Vos, W. M. The lactose transporter in *Leuconostoc lactis* is a new member of the LacS subfamily of galactoside-pentose-hexuronide translocators. *Appl. Environ. Microbiol.* **62:** 1574–1582, 1996.
120. Vaughan, E. E., David, S., Harrington, A., Daly, C., Fitzgerald, G. F., and de Vos, W. M. Characterization of plasmid-encoded citrate permease (citP) genes from *Leuconostoc* species reveals high sequence conservation with the *Lactococcus lactis* citP gene. *Appl. Environ. Microbiol.* **61:** 3172–3176, 1995.
121. Vescovo, M., Scolari, G. L., and Bottazzi, V. Plasmid-encoded ropiness production in *Lactobacillus casei* ssp. *casei. Biotechnol. Lett.* **11:** 709–712, 1989.
122. Vogel, R. F., Lohmann, M., Weller, A. N., Hugas, M., and Hammes, W. P. Structural similarity and distribution of small cryptic plasmids of *Lactobacillus curvatus* and *L. sake. FEMS Microbiol. Lett.* **84:** 183–190, 1991.
123. Volk, W. A., Bizzini, B., Jones, K. R., and Macrina, F. L. Inter- and intrageneric transfer of Tn916 between *Streptococcus faecalis* and *Clostridium tetani. Plasmid* **19:** 255–259, 1988.
124. Vujcic, M., and Topisirovic, L. Molecular analysis of the rolling-circle replicating plasmid pA1 of *Lactobacillus plantarum* A112. *Appl. Environ. Microbiol.* **59:** 274–280, 1993.
125. Weaver, K. E., and Clewell, D. B. Regulation of the pAD1 sex pheromone response in *Enterococcus faecalis*: Effects of host strain and traA, traB, and C region mutants on expression of an E region pheromone-inducible lacZ fusion. *J. Bacteriol.* **172:** 2633–2641, 1990.
126. Wright, G. D., Molinas, C., Arthur, M., Courvalin, P., and Walsh, C. T. Characterization of vanY, a DD-carboxypeptidase from vancomycin-resistant *Enterococcus faecium* BM4147. *Antimicrob. Agents Chemother.* **36:** 1514–1518, 1992.
127. Wyckoff, H. A., Barnes, M., Gillies, K. O., and Sandine, W. E. Characterization and sequence analysis of a stable cryptic plasmid from *Enterococcus faecium* 226 and development of a stable cloning vector. *Appl. Environ. Microbiol.* **62:** 1481–1486, 1996.
128. Yagi, Y., Kessler, R. E., Shaw, J. H., Lopatin, D. E., An, F., and Clewell, D. B. Plasmid content of *Streptococcus faecalis* strain 39-5 and identification of a pheromone (cPD1)-induced surface antigen. *J. Gen. Microbiol.* **129:** 1207–1215, 1983.
129. Yamamoto, N., and Takano, T. Isolation and characterization of a plasmid from *Lactobacillus helveticus* CP53. *Biosci. Biotechnol. Biochem.* **60:** 2069–2070, 1996.
130. Zink, A., Klein, J. R., and Plapp, R. Transformation of *Lactobacillus delbruckii* ssp.

lactis by electroporation and cloning of origin of replicon by use of a positive selection vector. *FEMS Microbiol. Lett.* **78:** 207–212, 1991.

131. Zuniga, M., Pardo, I., and Ferrer, S. Nucleotide sequence of plasmid p4028, a cryptic plasmid from *Leuconostoc oenos*. *Plasmid* **36:** 67–74, 1996.

3.3 VECTORS FOR LACTIC ACID BACTERIA

Vector	Replicon and Source[a]		Size (kb)	Marker[a]	Characteristics	Host[a]	Ref.
pEBM3		CO	9.6	Cm^r, Km^r		ESco, BFan	4
pECM3		CO	10.3	Cm^r, Km^r		ESco, BFan	4
pVS34	28Kb plasmid	LCld	8.1	Cm^r		LBpl, LCla, SAau	106
pVS40	28Kb plasmid	LCld	7.8	Nis^r		LBpl, LCla, SAau	106
pSC20	7Kb plasmid	LBpl	5.5	Cm^r, Em^r		ESco, BAsu, LBac, LBfe, LBhe, LBpl, LBre	25
pSC22	7Kb plasmid	LBpl	4.3	Cm^r, Em^r		ESco, BAsu, LBac, LBfe, LBhe, LBpl, LBre	25
pLPV111	p256	LBpl	4.2	Em^r	lacZ α-complementation	ESco, LBpl, LBsk	6, 53
pLPE317	p353-1	LBpe	2.9	Em^r		LBca, LBpe, LBpl	83
pLP3537	p353-2	LBpe	6.3	Em^r, Ap^r		ESco, LBac, LBca, LBpl, LBpe	24, 82, 83
pLPE323	p353-2	LBpe	3.6	Em^r		LBca, LBpe, LBpl	83
pLP825	p8014-2	LBpl	7.6	Cm^r, Ap^r		ESco, LBac, LBbr, LBca	23, 83
pH2311	pAMα1	ENfa	6.3	Em^r, Tc^r, Ap^r		ESco, LBca	91
pH2515	pAMα1	ENfa	6.6	Em^r, Tc^r, Ap^r		ESco, LBca	91
pHY300PLK	pAMα1	ENfa	4.9	Tc^r, Ap^r		ENfa, ESco, LBca	54, 77, 100, 108
pAMβ1	pAMβ1	ENfa	26.5	Em^r		LBac	68
pIL253	pAMβ1	ENfa	5.0	Em^r		BAsu, LBca, LBcu, LBsk, LBpl, LCla	42, 44, 55, 93
pIL::*nuc*MCS	pAMβ1	ENfa	5.8	Em^r, *nuc*	Insertional inactivation	ESco, LCla, STth	65
pTRK396	pAMβ1	ENfa	10.3	Em^r, *llaIR*	Positive selection	ESco, BAsu, CApi, Enfa, LBga, LBjo, LBpl, LCla	31
pTRK398	pAMβ1	ENfa	10.8	Em^r, *llaIR*	Positive selection	ESco, CApi, ENfa, LBjo, LCla	31
pTRKH2	pAMβ1	ENfa	6.9	Em^r	lacZ α-complementation	ESco, ENfa, LBjo, LCla	78

(*continued*)

Vector	Replicon and Source[a]		Size (kb)	Marker[a]	Characteristics	Host[a]	Ref.
pTRKL2	pAMβ1	ENfa	6.4	Em^r	lacZ α-complementation	ESco, ENfa, LBjo, LCla	78
pCP49	pBC1	BAco	7.0	Cm^r, Ap^r		ESco, BAam, BAsu, LBre, SAau, SAca	29a
pLM6	pBC1	BAco	2.8	Cm^r		BAam, BAsu, BAth, LBre, SAau, SAca	29a
pC194	pC194	SAau	2.9	Cm^r		BAsu, BAli, LBac, LBre	17, 68
pCI431	pCI411	LEla	5.8	Cm^r		ESco,BAsu, LBca, LCla, LEme, LEpa, STth	26
pCP53D	pCP53	LBhe	4.7	Tc^r		ESco, LBca, LBhe	111
p5aGFP2201a	pER8	STth	6.4	Em^r, Ap^r, *gfp*	Insertional inactivation	ESco, STth	96
pER82	pER8	STth	4.7	Cm^r, Em^r		STth	98
pMEGcat1	pER8	STth	7.4	Cm^r, Ap^r, *gfp*	Insertional inactivation	ESco, STth	96
pMEU5a	pER8	STth	5.5	Em^r, Ap^r		ESco, STth	95
pMEU6a	pER8	STth	5.5	Em^r, Ap^r		ESco, STth	95
pMEU9	pER8	STth	6.9	Cm^r, Em^r, Ap^r	Insertional inactivation	ESco, STth	95
pMEU10	pER8	STth	6.9	Cm^r, Em^r, Ap^r	Insertional inactivation	ESco, STth	95
pAM401	pGB354	STag	10.4	Cm^r, Tc^r		ESco, ENfa	109
pNCKH104	pGT232	LBre	5.7	Em^r		ESco, LBre	49
pGT633	pGT633	LBre	9.8	Em^r		BAsu, STsa, SAau, LBre, LBga, LBfe, LBsa, LBde	101
pGB301	pIP501	STag	9.8	Cm^r, Em^r		LBpl	8, 13
pGB354	pIP501	STag	6.3	Cm^r		LBac	12, 68
pSA3	pIP501	STag	10.2	Em^r, Cm^r, Tc^r		ESco, LBac, LBca, LBpl, LBre, LCla, STsa, STmu	10, 23, 28, 68, 76
pLUA105E	pLA105	LBac	7.8	Em^r, Ap^r		ESco, LBac, LBca	57, 58
pERM3.2	pLAB1000	LBhi	7.6	Em^r, Ap^r		LBpl, ESco	90
pLAB1102	pLAB1000	LBhi	7.5	Cm^r, Ap^r		ESco, BAsu, ENfa, LBca, LBpl	56
pLAB1301	pLAB1000	LBhi	5.2	Em^r, Ap^r		ESco, BAsu, ENfa, LBca, LBpl	37, 56
pLE16	pLB10	LBde	7.6	Cm^r, Tc^r		LBsp, ESco	22
pJK352	pLC2	LBcu	5.9	Cm^r, Ap^r		ESco, BAsu, LBca, LCla	59
pJK355	pLC2	LBcu	3.2	Cm^r		BAsu, LBca, LCla	59
pJK356	pLC2	LBcu	3.2	Cm^r		BAsu, LBca, LCla	59
pLEM415	pLEM3	LBfe	6.3	Em^r, Ap^r		ESco, LBfe	39
pLEM7	pLEM3	LBfe	3.5	Em^r		LBfe	39
pLHR	pLJ1	LBhe	8.5	Em^r, Ap^r		LBhe, ESco,	46
pULP8	pLP1	LBpl	6.6	Em^r, Ap^r		ESco, BAsu, LBpl	18

(*continued*)

Vector	Replicon and Source[a]		Size (kb)	Marker[a]	Characteristics	Host[a]	Ref.
pULP9	pLP1	LBpl	6.8	Em^r, Ap^r		ESco, BAsu, LBpl	18
pLUL200	pLUL631	LBre	6.0	Cm^r		ESco, LBre,	2
pDG7	pMB1	BFlo	7.3	Ap^r, Cm^r		ESco, BFlo, BFan, BFbr, BFbi, BFin	4, 73, 85
pDGE7	pMB1	BFlo		Cm^r, Em^r		ESco, BFan	85
pNC7	pMB1	BFlo	4.9	Cm^r		BFan	85
pRM2	pMB1	BFlo	6.0	Sp^r, Ap^r		ESco, BFlo	74
pHW800	pMBB1	ENfm	3.8	Cm^r		LC, LE, PD	110
pLSE1	pMV158	STag	5.6	Em^r, Tc^r		ESco, STor, STpn	30
pAM610	pMV163	STag	9.5	Km^r, Tc^r		ESco, ENfa	9
pAZ20	pNCDO151	LBca	8.3	Cm^r, Ap^r		ESco, LBde	113
pTRK13	pPM4	LBac	12.5	Cm^r		ENfa, LBac	68
pDBN183	pSH71	LCla	8.8	Cm^r, Km^r, Ap^r		ESco, LBca, STth	97
pNZ12	pSH71	LCla	4.3	Cm^r, Km^r		ESco, LBca, LBcu, LBpl, LBsk, LEpa	23, 29
pNZ123	pSH71	LCla	2.8	Cm^r		ESco, LBac, LBca, LBpl	81
pNZ124	pSH71	LCla	2.8	Cm^r		ESco, STth	19
pNZ17	pSH71	LCla	5.7	Cm^r, Km^r		ESco, LBfe, LBac	87
pNZ280	pSH71	LCla	4.7	Cm^r, Tc^r		ESco, BAsu, LCla, LBpl	80
pVS2	pSH71	LCla	5.0	Cm^r, Em^r		ESco, BAsu, LBpl, LBsk, LCla	7, 92
pDB101	pSM19035	STpy	17.8	Em^r		BAsu, STsa	11, 16
pHPS9	pTA1060	BAsu	5.7	Cm^r, Em^r		ESco, BAsu, PDac	20, 45
pBLES100	pTB6	BFlo	9.1	Sp^r		ESco, BFlo	72
pDL276	pVA380-1	STfe	6.9	Km^r		ESco, STmu, STgo, ENfa	34, 52
pDL278	pVA380-1	STfe	6.6	Sp^r		STc^r, STsa, STgo, STor	27, 34
pDL412	pVA380-1	STfe	6.2	Sm^r		STsa	62
pDL413	pVA380-1	STfe	5.7	Km^r		STsa	62
pDL414	pVA380-1	STfe	5.7	Em^r		STsa	62
pDL421	pVA380-1	STfe	14	Tc^r		STsa	63
pVA749	pVA380-1	STfe	5.2	Em^r		ESco, STsa	99
pVA797	pVA380-1	STfe	30.7	Cm^r		LBac, STsa	36, 67, 68
pVA838	pVA380-1	STfe	9.2	Cm^r, Em^r		ESco, STsa	71
pVA856	pVA380-1	STfe	9.2	Cm^r, Em^r, Tc^r		ESco, STsa	70
pCP12	pWC1	LCla	3.9	Cm^r		STth, ENfa, SAau	79
pJK300	pWS97	LBde	6.8	Cm^r, Ap^r		ESco, LBde	113
pGK12	pWV01	LCla	4.4	Cm^r, Em^r		ESco, LCla, LBac, LBca, LBde, LBfe, LBpl, LBre, LBhe, LEde, LEla, PDac	14, 60, 68, 69, 84, 112, 113
pGKV1	pWV01	LCla	4.6	Cm^r, Em^r		LBac	68
pGKV13	pWV01	LCla	5.0	Cm^r, Em^r		ESco, LCla, LBpl	8, 38
pGKV2	pWV01	LCla	4.6	Cm^r, Em^r		ESco, LCla, LBpl	56

(continued)

Vector	Replicon and Source[a]		Size (kb)	Marker[a]	Characteristics	Host[a]	Ref.
pGKV21	pWV01	LCla	4.7	Cm^r, Em^r		ESco, LBpl, LCla	64, 103,
pGK::*nuc*MCS	pWV01	LCla	4.3	Em^r, *nuc*	Insertional inactivation	ESco, LCla, STth	65
pTRK434	pWV01	LCla	5.5	Cm^r, Em^r, *lafI*		ESco, LBac, LBfe, LBga, LBjo	3

Vector	Replicon and Source[a]		Size (kb)	Marker[a]	Reporter Gene[a]	Host[a]	Ref.
Promoter Screening Vector							
pNCKH113	pGT232	LBre	7.0	Em^r	*bglM*	ESco, LBre	49
pLSE4	pMV158	STag	6.6	Em^r, Tc^r	*litA*	ESco, STor, STpn	30
pGIP331	pSH71	LCla	8.3	Cm^r, Km^r, Sm^r	*amyL*	ESco, ENfa, LBpl	50, 51
pNZ272	pSH71	LCla	4.7	Cm^r	*gusA*	ESco, LBca, LBpl, LCla, LEla	81
pMU1327	pVA380-1	STfe	7.5	Em^r	*cat*	ESco, STsa	1
pMU1328	pVA380-1	STfe	7.5	Em^r	*cat*	ESco, STsa, LCla	1, 61
pTG244	pVA380-1	STfe	9.5	Em^r, Ap^r	*cat*-86	ESco, LCla, STth	94
pBV5030	pWV01	LCla	4.4	Em^r	*cat*-86	ESco, LBre, LCla	15, 31, 33
pGKV210	pWV01	LCla	4.5	Em^r	*cat*-86	ESco, BAsu, LBca, LBjo, LBre, LCla, LEla, STth	32, 40, 41, 43, 104
Promoter-Secretion Signal Screening Vector							
pGIP212	pWV01	LCla	8.2	Cm^r, Em^r, Sm^r	*amyL*	ESco, ENfa, LBpl	50, 51

Vector	Replicon and Source[a]		Size (kb)	Marker[a]	Region Involved in Integration	Recipient[a]	Ref.
Integration Vector							
pTKR182	pACYC184	ESco	10.5	Cm^r, Em^r	attP, int (phage ϕadh)	LBga	84
pTRK327	pACYC184	ESco	8.2	Cm^r, Em^r, Tc^r	IS*1223*	LBga	107
pCI192	pBR322	ESco	9.0	Cm^r, Tc^r	Tn*919*-derived fragment	ENfa, LCla	21
pSATE253	pBR322	ESco	11.9	Em^r, *amy*, *celA*	Chromosome DNA (LBpl)	LBpl	90
pMYL1-βgal	pBluescript II SK$^+$	ESco	9.1	Ap^r, *lacZ*	Chromosome DNA (LBac)	LBac	66
pSF28	pCI305	LCla	8.3	Cm^r	attP, int (phage ϕSfi21)	STth	19
pMC1	pRC1	ESco	5.2	Em^r	attP, int (phage mv4)	LBpl, LBca, LCla, ENfa, STpn	5, 35
pAM6200	pUC19	ESco	3.4	Sp^r	Conserved region of recA (STgo)	STgo	105
pDP225	pUC19	ESco	8.7	Ap^r	*lacZ* with deletion (STth)	STth	75

(*continued*)

Vector	Replicon and Source[a]		Size (kb)	Marker[a]	Promoter/Secretion Signal[a]	Host[a]	Ref.
Expression/Secretion Vector							
pER82Pb	pER8	STth	4.8	Cm^r, Em^r	sP1 promoter	STth	98
pBG10	pLJ1	LBhe	6.0	*lacZ*	*erm* promoter	ESco, LBhe	47
pCMR100	pSH71	LCla	6.2	Cm^r, Km^r	PA promoter/ PA secretion signal	ESco, LBfe, LBpl, LBze	88
pCMR110	pSH71	LCla	3.5	Cm^r	PA promoter/ PA secretion signal	ESco, LBfe, LBpl, LBze	88
pKTH2121	pWV01	LCla	6.0	Cm^r, Em^r	*slpA* promoter/ *slpA* secretion signal	ESco, LBbr, LBca, LBga, LBpl, LCla	89
pMG36e	pWV01	LCla	3.7	Em^r	P32 promoter	ESco, LBga, LCla	86, 102

Vector	Replicon and Source[a]		Size (kb)	Marker[a]	Host[a]	Ref.
Replication Origin Screening Vector						
pCI341	pBR322	ESco	3.1	Cm^r	LCla, LEla	26, 48
pGI4010	pBR322	ESco	3.7	Em^r, Ap^r	LBpl	56
pAZ8	pUH84	ESco	4.5	Cm^r, Ap^r	LBde	113

[a]Abbreviations
amy: α-amylase gene,
Ap^r : ampicillin resistance,
BAam: *Bacillus amyloliquefaciens,*
BAco: *Bacillus coagulans,*
BAli: *Bacillus licheniformis,*
BAsu: *Bacillus subtilis,*
BAth: *Bacillus huringiensis,*
BFan: *B. animalis,*
BFbi: *B. bifidum,*
BFbr: *B. breve,*
BFin: *B. infantis,*
BFlo: *B. longum,*
bglM: β-glucanase gene,
CApi: *Carnobacterium pisicola,*
cat: chloramphenicol acetyltransferase gene,
celA: endoglucanase,
Cm^r: chloramphenicol resistance,
CO: *Colynebacterium,*
Em^r: erythromycin resistance,
ENfa: *E. faecalis,*
ENfm: *E. faecium,*
erm: erythromycin resistance gene,
ESco: *Escherichia coli,*
gfp: green fluorescent protein gene,
gusA: β-glucuronidase gene,
Km^r: kanamycin resistance,
lacZ: β-galactosidase gene,
lafI: lactacin F immunity,
LBac: *L. acidophilus,*
LBbr: *L. brevis,*
LBca: *L. casei,*
LBcu: *L. curvatus,*
LBde: *L. delbrueckii,*
LBfe: *L. fermentum,*
LBga: *L. gasseri,*
LBhe: *L. helveticus,*
LBhi: *L. hilgardii,*
LBjo: *L. johnsonii,*
LBpe: *L. pentosus,*
LBpl: *L. plantarum,*
LBre: *L. reuteri,*
LBsa: *L. salivarius,*
LBsk: *L. sake,*
LBsp: *Lactobacillus* sp.,
LBze: *L. zeae,*
LC: *Lactococcus,*
LCla: *Lactococcus lactis,*
LCld: *Lactococcus lactis* subsp. *lactis* biovar *diacetylactis,*
LE: *Leuconostoc,*
LEde: *Leuconostoc dextranicum,*
LEla: *Leuconostoc lactis,*
LEme: *Leuconostoc mesenteroides,*
LEpa: *Leuconostoc paramesenteroides,*
litA: *N*-acetyl-muramyl-L-alanine amidase gene,
llaIR: restriction endonuclease cassette,
Nis^r: nisin resistance, *nuc*: nuclease gene,
PA: protein A,
PD: *Pediococcus,*
PDac: *P. acidilactici,*
SAau: *Staphylococcus aureus,*
SAca: *Staphylococcus carnosus,*
slpA: S-layer protein gene,
Sm^r: streptomycin resistance,
Sp^r: spectinomycin resistance,
STag: *S. agalactiae,*
STcr: *S. crista,*
STfe: *S. ferus,*
STgo: *S. gordonii,*
STmu: *S. mutans,* STor: *S. oralis,*
STpn: *S. pneumoniae,*
STpy: *S. pyogenes,* STsa: *S. sanguis,*
STth: *S. thermophilus,*
Tc^r: tetracycline resistance.

REFERENCES

1. Achen, M. G., Davidson, B. E., and Hillier, A. J. Construction of plasmid vectors for the detection of streptococcal promoters. *Gene* **45:** 45–49, 1986.
2. Ahrne, S., Molin, G., and Axelsson, L. Transformation of *Lactobacillus reuteri* with electroporation: Studies on the erythromycin resistance plasmid pLUL631. *Curr. Microbiol.* **24:** 199–205, 1992.
3. Allison, G. E., and Klaenhammer, T. R. Functional analysis of the gene encoding immunity to lactacin F lafI and its use as a *Lactobacillus*-specific food-grade genetic marker. *Appl. Environ. Microbiol.* **62:** 4450–4460, 1996.
4. Argnani, A. Leer, R. J., van Luijk, N., and Pouwels, P. H. A convenient and reproducible method to genetically transform bacteria of the genus *Bifidobacterium. Microbiology* **142:** 109–114, 1996.
5. Auvray, F., Coddeville, M., Ritzenthaler, P., and Dupont, L. Plasmid integration in a wide range of bacteria mediated by the integrase of *Lactobacillus delbrueckii* bacteriophage mv4. *J. Bacteriol.* **179:** 1837–1845, 1997.
6. Axelsson, L., and Holck, A. The genes involved in the production of and Immunity to sakacin A a bacteriocin from *Lactobacillus sake* Lb706. *J. Bacteriol.* **177:** 2125–2137, 1995.
7. Axelsson, L., Holck, A., Birkeland, S. E., Aukrust, T., and Blom, H. Cloning and nucleotide sequence of a gene from *Lactobacillus sake* Lb706 necessary for sakacin A production and immunity. *Appl. Environ. Microbiol.* **59:** 2868–2875, 1993.
8. Badii, R., Jones, S., and Warner, P. J. Spheroplast and electroporation-mediated transformation of *Lactobacillus plantarum. Lett. Appl. Microbiol.* **9:** 41–44, 1989.
9. Baik, B., and Pack, M. Y. Expression of *Bacillus subtilis* endoglucanase gene in *Lactobacillus acidophilus. Biotechnol. Lett.* **12:** 919–924, 1990.
10. Bates, E. E., Gilbert, H. J., Hazlewood, G.P,. Huckle, J., Laurie, J. I., and Mann, S. P. Expression of a *Clostridium thermocellum* endoglucanase gene in *Lactobacillus plantarum. Appl. Environ. Microbiol.* **55:** 2095–2097, 1989.
11. Behnke, D., and Ferretti, J. J. Physical mapping of plasmid pDB101: A potential vector plasmid for molecular cloning in streptococci. *Plasmid* **4:** 130–138, 1980.
12. Behnke, D., and Gilmore, M. S. Location of antibiotic resistance determinants copy control and replication functions on the double-selective streptococcal cloning vector pGB301. *Mol. Gen. Genet.* **184:** 115–120, 1981.
13. Behnke, D., Gilmore, M. S., and Ferretti, J. J. Plasmid pGB301 a new multiple resistance streptococcal cloning vehicle and its use in cloning of a gentamicin/kanamycin resistance determinant. *Mol. Gen. Genet.* **182:** 414–421, 1981.
14. Bhowmik, T., and Steele, J. L. Cloning characterization and insertional inactivation of the *Lactobacillus helveticus* D(-) lactate dehydrogenase gene. *Appl. Microbiol. Biotechnol.* **41:** 432–439, 1994.
15. Bojovic, B., Djordjevic, G., and Topisirovic, L. Improved vector for promoter screening in lactococci. *Appl. Environ. Microbiol.* **57:** 385–388, 1991.
16. Breitling, R., Gerlach, D., Hartmann, M., and Behnke, D. Secretory expression in *Escherichia coli* and *Bacillus subtilis* of human interferon alpha genes directed by staphylokinase signals. *Mol. Gen. Genet.* **217:** 384–391, 1989.
17. Brigidi, P., Bolognani, F., Rossi, M., Cerre, C., and Matteuzzi, D. Cloning of the gene

for cholesterol oxidase in *Bacillus* spp. *Lactobacillus reuteri* and its expression in *Escherichia coli. Lett. Appl. Microbiol.* **17:** 61–64, 1993.

18. Bringel, F., Frey, L., and Hubert, J. C. Characterization cloning curing and distribution in lactic acid bacteria of pLP1 a plasmid from *Lactobacillus plantarum* CCM 1904 and its use in shuttle vector construction. *Plasmid* **22:** 193–202, 1989.
19. Bruttin, A., Foley, S., and Brussow, H. The site-specific integration system of the temperate *Streptococcus thermophilus* bacteriophage phiSfi21. *Virology* **237:** 148–158, 1997.
20. Bukhtiyarova, M., Yang, R., and Ray, B. Analysis of the pediocin AcH gene cluster from plasmid pSMB74 and its expression in a pediocin-negative *Pediococcus acidilactici* strain. *Appl. Environ. Microbiol.* **60:** 3405–3408, 1994.
21. Casey, J., Daly, C., and Fitzgerald, G. F. Chromosomal integration of plasmid DNA by homologous recombination in *Enterococcus faecalis* and *Lactococcus lactis* subsp. *lactis* hosts harboring Tn919. *Appl. Environ. Microbiol.* **57:** 2677–2682, 1991.
22. Chagnaud, P., Chan, K. C., Duran, R., Naouri, P., Arnaud, A., and Galzy, P. Construction of a new shuttle vector for *Lactobacillus. Can. J. Microbiol.* **38:** 69–74, 1992.
23. Chassy, B. M., and Flickinger, J. L. Transformation of *Lactobacillus casei* by electroporation. *FEMS Microbiol. Lett.* **44:** 173–177, 1987.
24. Christiaens, H., Leer, R. J., Pouwels, P. H., and Verstraete, W. Cloning and expression of a conjugated bile acid hydrolase gene from *Lactobacillus plantarum* by using a direct plate assay. *Appl. Environ. Microbiol.* **58:** 3792–3798, 1992.
25. Cocconcelli, P. S., Gasson, M. J., Morelli, L., and Bottazzi, V. Single-stranded DNA plasmid vector construction and cloning of *Bacillus stearothermophilus* α-amylase in *Lactobacillus. Res. Microbiol.* **142:** 643–652, 1991.
26. Coffey, A., Harrington, A., Kearney, K., Daly, C., and Fitzgerald, G. Nucleotide sequence and structural organization of the small broad-host-range plasmid pCI411 from *Leuconostoc lactis* 533. *Microbiology* **140:** 2263–2269, 1994.
27. Correia, F. F., McKay, T. L., Farrow, M. F., Rosan, B., and DiRienzo, J. M. Natural transformation of *Streptococcus crista. FEMS Microbiol. Lett.* **143:** 13–18, 1996.
28. Dao, M. L., and Ferretti, J. J. *Streptococcus-Escherichia coli* shuttle vector pSA3 and its use in the cloning of streptococcal genes. *Appl. Environ. Microbiol.* **49:** 115–119, 1985.
29. David, S., Simons, G., and de Vos, W. M. Plasmid transformation by electroporation of *Leuconostoc paramesenteroides* and its use in molecular cloning. *Appl. Environ. Microbiol.* **55:** 1483–1489, 1989.

29a. de Rossi, E., Brigidi, P., Rossi, M., Matteuzzi, D., and Riccardi, G. Characterization of gram-positive broad host-range plasmids carrying a thermophilic replicon. *Res. Microbiol.* **142:** 389–396, 1991.

30. Diaz, E., and Garcia, J. L. Construction of a broad-host-range pneumococcal promoter-probe plasmid. *Gene* **90:** 163–167, 1990.
31. Djordjevic, G. M., and Klaenhammer, T. R. Positive selection cloning vectors for gram-positive bacteria based on a restriction endonuclease cassette. *Plasmid* **35:** 37–45, 1996.
32. Djordjevic, G. M., Bojovic, B., Banina, A., and Topisirovic, L. Cloning of promoter-like sequences from *Lactobacillus paracasei* subsp. *paracasei* CG11 and their expression in *Escherichia coli Lactococcus lactis* and *Lactobacillus reuteri. Can. J. Microbiol.* **40:** 1043–1050, 1994.

33. Djordjevic, G. M., Bojovic, B., Miladinov, N., and Topisirovic, L. Cloning and molecular analysis of promoter-like sequences isolated from the chromosomal DNA of *Lactobacillus acidophilus* ATCC 4356. *Can. J. Microbiol.* **43:** 61–69, 1997.
34. Dunny, G. M., Lee, L. N., and LeBlanc, D. J. Improved electroporation and cloning vector system for gram-positive bacteria. *Appl. Environ. Microbiol.* **57:** 1194–1201, 1991.
35. Dupont, L., Boizet-Bonhoure, B., Coddeville, M., Auvray, F., and Ritzenthaler, P. Characterization of genetic elements required for site-specific integration of *Lactobacillus delbrueckii* subsp. *bulgaricus* bacteriophage mv4 and construction of an integration-proficient vector for *Lactobacillus plantarum. J. Bacteriol.* **177:** 586–595, 1995.
36. Evans, R. P. Jr., and Macrina, F. L. Streptococcal R plasmid pIP501: Endonuclease site map resistance determinant location and construction of novel derivatives. *J. Bacteriol.* **154:** 1347–1355, 1983.
37. Ferain, T., Garmyn, D., Bernard, N., Hols, P., and Delcour, J. *Lactobacillus plantarum ldhL* gene: Overexpression and deletion. *J. Bacteriol.* **176:** 596–601, 1994.
38. Fitzsimons, A., Hols, P., Jore, J., Leer, R. J., O'Connell, M., and Delcour, J. Development of an amylolytic *Lactobacillus plantarum* silage strain expressing the *Lactobacillus amylovorus* α-amylase gene. *Appl. Environ. Microbiol.* **60:** 3529–3535, 1994.
39. Fons, M., Hege, T., Ladire, M., Raibaud, P., Ducluzeau, R., and Maguin, E. Isolation and characterization of a plasmid from *Lactobacillus fermentum* conferring erythromycin resistance. *Plasmid* **37:** 199–203, 1997.
40. Fremaux, C., de Antoni, G. L., Raya, R. R., and Klaenhammer, T. R. Genetic organization and sequence of the region encoding integrative functions from *Lactobaillus gasseri* temperate bacteriophage ϕadh. *Gene* **126:** 61–66, 1993.
41. Fremaux, C., Hechard, Y., and Cenatiempo, Y. Mesentericin Y105 gene clusters in *Leuconostoc mesenteroides* Y105. *Microbiology* **141:** 1637–1645, 1995.
42. Gaier, W., Vogel, R. F., and Hammes, W. P. Genetic transformation of intact cells of *Lactobacillus curvatus* Lc2-c and *Lact. sake* Ls2 by electroporation. *Lett. Appl. Microbiol.* **11:** 81–83, 1990.
43. Gaier, W. Vogel, R. F., and Hammes, W. P. Cloning and expression of the lysostaphin gene in *Bacillus subtilis* and *Lactobacillus casei. Lett. Appl. Microbiol.* **14:** 72–76, 1992.
44. Gold, R. S., Meagher, M. M., Tong, S., Hutkins, R. W., and Conway T. Cloning and expression of the *Zymomonas mobilis* "production of ethanol" genes in *Lactobacillus casei. Curr. Microbiol.* **33:** 256–260, 1996.
45. Haima, P., van Sinderen, D., Schotting, H., Bron, S., and Venema, G. Development of a β-galactosidase α-complementation system for molecular cloning in *Bacillus subtilis. Gene* **86:** 63–69, 1990.
46. Hashiba, H., Takiguchi, R., Ishii, S., and Aoyama, K. Transformation of *Lactobacillus helveticus* subsp. *jugurti* with plasmid pLHR by electroporation. *Agric. Biol. Chem.* **54:** 1537–1541, 1990.
47. Hashiba, H., Takiguchi, R., Jyoho, K., and Aoyama, K. Establishment of a host-vector system in *Lactobacillus helveticus* with β-galactosidase activity as a selection marker. *Biosci. Biotechnol. Biochem.* **56:** 190–194, 1992.
48. Hayes, F., Daly, C., and Fitzgerald, G. F. Identification of the minimal replicon of *Lactococcus lactis* subsp. *lactis* UC317 plasmid pCI305. *Appl. Environ. Microbiol.* **56:** 202–209, 1990.

49. Heng, N. C., Jenkinson, H. F., and Tannock, G. W. Cloning and expression of an endo-1 3-1 4-β-glucanase gene from *Bacillus macerans* in *Lactobacillus reuteri*. *Appl. Environ. Microbiol.* **63:** 3336–3340, 1997.

50. Hols, P., Baulard, A., Garmyn, D., Delplace, B., Hogan, S., and Delcour, J. Isolation and characterization of genetic expression and secretion signals from *Enterococcus faecalis* through the use of broad-host-range α-amylase probe vectors. *Gene* **118:** 21–30, 1992.

51. Hols, P., Ferain, T., Garmyn, D., Bernard, N., and Delcour, J. Use of homologous expression-secretion signals and vector-free stable chromosomal integration in engineering of *Lactobacillus plantarum* for α-amylase and levanase expression. *Appl. Environ. Microbiol.* **60:** 1401–1413, 1994.

52. Homonylo-McGavin, M. K., and Lee, S. F. Role of the C terminus in antigen P1 surface localization in *Streptococcus mutans* and two related cocci. *J. Bacteriol.* **178:** 801–807, 1996.

53. Huhne, K., Axelsson, L., Holck, A., and Krockel, L. Analysis of the sakacin P gene cluster from *Lactobacillus sake* Lb674 and its expression in sakacin-negative *Lb. sake* strains. *Microbiology* **142:** 1437–1448, 1996.

54. Ishiwa, H., and Shibahara-Sone, H. New shuttle vectors for *Escherichia coli* and *Bacillus subtilis*. IV. The nucleotide sequence of pHY300PLK and some properties in relation to transformation. *Jpn. J. Genet.* **61:** 515–528, 1986.

55. Jones, S., and Warner, P. J. Cloning and expression of α-amylase from Bacillus amyloliquefaciens in a stable plasmid vector in *Lactobacillus plantarum*. *Lett. Appl. Microbiol.* **11:** 214–219, 1990.

56. Josson, K., Scheirlinck, T., Michiels, F., Platteeuw, C., Stanssens, P., Joos, H., Dhaese, P., Zabeau, M., and Mahillon, J. Characterization of a gram-positive broad-host-range plasmid isolated from *Lactobacillus hilgardii*. *Plasmid* **21:** 9–20, 1989.

57. Kanatani, K., Oshimura, M., and Sano, K. Isolation and characterization of acidocin A and cloning of the bacteriocin gene from *Lactobacillus acidophilus*. *Appl. Environ. Microbiol.* **61:** 1061–1067, 1995.

58. Kanatani, K., Yoshida, K., Tahara, T., Yamada, K., Miura, H., Sakamoto, M., and Oshimura, M. Transformation of *Lactobacillus acidophilus* TK1892 by electroporation with pULA105E plasmid. *J. Ferment. Bioeng.* **74:** 358–362, 1992.

59. Klein, J. R., Ulrich, C., and Plapp, R. Characterization and sequence analysis of a small cryptic plasmid from *Lactobacillus curvatus* LTH683 and its use for construction of new *Lactobacillus* cloning vectors. *Plasmid* **30:** 14–29, 1993.

60. Kok, J., van der Vossen, J. M., and Venema, G. Construction of plasmid cloning vectors for lactic streptococci which also replicate in *Bacillus subtilis* and *Escherichia coli*. *Appl. Environ. Microbiol.* **48:** 726–731, 1984.

61. Lakshmidevi, G., Davidson, B. E., and Hillier, A. J. Molecular characterization of promoters of the *Lactococcus lactis* subsp. *cremoris* temperate bacteriophage BK5-T and identification of a phage gene implicated in the regulation of promoter activity. *Appl. Environ. Microbiol.* **56:** 934–942, 1990.

62. LeBlanc, D. J., Inamine, J. M., and Lee, L. N. Broad geographical distribution of homologous erythromycin kanamycin and streptomycin resistance determinants among group D streptococci of human and animal origin. *Antimicrob. Agents Chemother.* **29:** 549–555, 1986.

63. LeBlanc, D. J., Lee, L. N., Titmas, B. M., Smith, C. J., and Tenover, F. C. Nucleotide se-

quence analysis of tetracycline resistance gene *tetO* from *Streptococcus mutans* DL5. *J. Bacteriol.* **170:** 3618–3626, 1988.

64. Leer, R. J., van der Vossen, J. M., van Giezen, M., van Noort, J. M., and Pouwels, P. H. Genetic analysis of acidocin B a novel bacteriocin produced by *Lactobacillus acidophilus. Microbiology* **141:** 1629–1635, 1995.
65. Le Loir, Y., Gruss, A., Ehrlich, S. D., and Langella, P. Direct screening of recombinants in gram-positive bacteria using the secreted staphylococcal nuclease as a reporter. *J. Bacteriol.* **176:** 5135–5139, 1994.
66. Lin, M. Y., Harlander, S., and Savaiano, D. Construction of an integrative food-grade cloning vector for *Lactobacillus acidophilus. Appl. Microbiol. Biotechnol.* **45:** 484–489, 1996.
67. Luchansky, J. B., Kleeman, E. G., Raya, R. R., and Klaenhammer, T. R. Genetic transfer systems for delivery of plasmid deoxyribonucleic acid to *Lactobacillus acidophilus* ADH: Conjugation electroporation and transduction. *J. Dairy Sci.* **72:** 1408–1417, 1989.
68. Luchansky, J. B., Muriana, P. M., and Klaenhammer, T. R. Application of electroporation for transfer of plasmid DNA to *Lactobacillus Lactococcus Leuconostoc Listeria Pediococcus Bacillus Staphylococcus Enterococcus* and *Propionibacterium. Mol. Microbiol.* **2:** 637–646, 1988.
69. Luchansky, J. B. Tennant, M. C., and Klaenhammer, T. R. Molecular cloning and deoxyribonucleic acid polymorphisms in *Lactobacillus acidophilus* and *Lactobacillus gasseri. J. Dairy Sci.* **74:** 3293–3302, 1991.
70. Macrina, F. L., Evans, R. P., Tobian, J. A., Hartley, D. L., Clewell, D. B., and Jones, K. R. Novel shuttle plasmid vehicles for *Escherichia-Streptococcus* transgeneric cloning. *Gene* **25:** 145–150, 1983.
71. Macrina, F. L., Tobian, J. A., Jones, K. R.,Evans, R. P., and Clewell, D. B. A cloning vector able to replicate in *Escherichia coli* and *Streptococcus sanguis. Gene* **19:** 345–353, 1982.
72. Matsumura H., Takeuchi, A., and Kano, Y. Construction of *Escherichia coli-Bifidobacterium longum* shuttle vector transforming *B. longum* 105-A and 108-A. *Biosci. Biotechnol. Biochem.* **61:** 1211–1212, 1997.
73. Matteuzzi, D., Brigidi, P., Rossi, M., and Di, D. Characterization and molecular cloning of *Bifidobacterium longum* cryptic plasmid pMB1. *Lett. Appl. Microbiol.* **11:** 220–223, 1990.
74. Missich, R., Sgorbati, B., and LeBlanc, D. J. Transformation of *Bifidobacterium longum* with pRM2 a constructed *Escherichia coli—B. longum* shuttle vector. *Plasmid* **32:** 208–211, 1994.
75. Mollet, B., Knol ,J., Poolman, B., Marciset, O., and Delley, M. Directed genomic integration gene replacement and integrative gene expression in *Streptococcus thermophilus. J. Bacteriol.* **175:** 4315–4324, 1993.
76. Morelli, L., Cocconcelli, P. S., Bottazzi, V., Damiani, G., Ferretti, L, and Sgaramella, V. *Lactobacillus* protoplast transformation. *Plasmid* **17:** 73–75, 1987.
77 Natori, Y., Kano, Y., and Imamoto, F. Genetic transformation of *Lactobacillus casei* by electroporation. *Biochimie* **72:** 265–269, 1990.
78. O'Sullivan, D. J., and Klaenhammer, T. R. High- and low-copy-number *Lactococcus* shuttle cloning vectors with features for clone screening. *Gene* **137:** 227–231, 1993.

79. Pillidge, C. J., Cambourn, W. M., and Pearce, L. E. Nucleotide sequence and analysis of pWC1 a pC194-type rolling circle replicon in *Lactococcus lactis. Plasmid* **35:** 131–140, 1996.

80. Platteeuw, C., Michiels, F., Joos, H., Seurinck, J., and de Vos, W. M. Characterization and heterologous expression of the tetL gene and identification of iso-ISS1 elements from *Enterococcus faecalis* plasmid pJH1. *Gene* **160:** 89–93, 1995.

81. Platteeuw, C., Simons, G., and de Vos, W. M. Use of the *Escherichia coli* β-glucuronidase (*gusA*) gene as a reporter gene for analyzing promoters in lactic acid bacteria. *Appl. Environ. Microbiol.* **60:** 587–593, 1994.

82. Posno, M., Heuvelmans, P. T., van Giezen, M. J., Lokman, B. C., Leer, R. J., and Pouwels, P. H. Complementation of the inability of *Lactobacillus* strains to utilize D-xylose with D-xylose catabolism-encoding genes of *Lactobacillus pentosus. Appl. Environ. Microbiol.* **57:** 2764–2766, 1991.

83. Posno, M., Leer, R. J., van Luijk, N., van Giezen, M. J. F., Heuvelmans, P. T. H. M., Lokman, B. C., and Pouwels, P. H. Incompatibility of *Lactobacillus* vectors with replicons derived from small cryptic *Lactobacillus* plasmids and segregational instability of the introduced vectors. *Appl. Environ. Microbiol.* **57:** 1822–1828, 1991.

84. Raya R. R. Fremaux C. de Antoni G. L., and Klaenhammer T. R. Site-specific integration of the temperate bacteriophage ϕadh into the *Lactobacillus gasseri* chromosome and molecular characterization of the phage (attP) and bacterial (attB) attachment sites. *J. Bacteriol.* **174:** 5584–5592, 1992.

85. Rossi, M., Brigidi, P., Gonzalez-Vara-y-Rodriguez, A., and Matteuzzi, D. Characterization of the plasmid pMB1 from *Bifidobacterium longum* and its use for shuttle vector construction. *Res. Microbiol.* **147:** 133–143, 1996.

86. Roy, D. G., Klaenhammer, T. R., and Hassan, H. M. Cloning and expression of the manganese superoxide dismutase gene of *Escherichia coli* in *Lactococcus lactis* and *Lactobacillus gasseri. Mol. Gen. Genet.* **239:** 33–40, 1993.

87. Rush, C. M., Hafner, L. M., and Timms, P. *Lactobacilli*: Vehicles for antigen delivery to the female urogenital tract. *Adv. Exp. Med. Biol.* **371B:** 1547–1552, 1995.

88. Rush, C. M., Hafner, L. M., and Timms, P. Protein A as a fusion partner for the expression of heterologous proteins in *Lactobacillus. Appl. Microbiol. Biotechnol.* **47:** 537–542, 1997.

89. Savijoki, K., Kahala, M., and Palva, A. High level heterologous protein production in *Lactococcus* and *Lactobacillus* using a new secretion system based on the *Lactobacillus brevis* S-layer signals. *Gene* **186:** 255–262, 1997.

90. Scheirlinck, T., Mahillon, J., Joos, H., Dhaese, P., and Michiels, F. Integration and expression of α-amylase and endoglucanase genes in the *Lactobacillus plantarum* chromosome. *Appl. Environ. Microbiol.* **55:** 2130–2137, 1989.

91. Shimizu-Kadota, M., Shibahara-Sone, H., and Ishiwa, H. Shuttle plasmid vectors for *Lactobacillus casei* and *Escherichia coli* with a minus origin. *Appl. Environ. Microbiol.* **57:** 3292–3300, 1991.

92. Sibakov, M., Koivula, T., von Wright, A., and Palva, I. Secretion of TEM β-lactamase with signal sequences isolated from the chromosome of *Lactococcus lactis* subsp. *lactis. Appl. Environ. Microbiol.* **57:** 341–348, 1991.

93. Simon, D., and Chopin, A. Construction of a vector plasmid family and its use for molecular cloning in *Streptococcus lactis. Biochimie* **70:** 559–566, 1988.

94. Slos, P,. Bourquin, J. C. Lemoine, Y., and Mercenier, A. Isolation and characterization of chromosomal promoters of *Streptococcus salivarius* subsp. *thermophilus*. *Appl. Environ. Microbiol.* **57:** 1333–1339, 1991.
95. Solaiman, D. K., and Somkuti, G. A. Shuttle vectors developed from *Streptococcus thermophilus* native plasmid. *Plasmid* **30:** 67–78, 1993.
96. Solaiman, D. K. Y., and Somkuti, G. A. Construction of a green-fluorescent protein-based insertion-inactivation shuttle vector for lactic acid bacteria and *Escherichia coli*. *Biotechnol. Lett.* **19:** 1175–1179, 1997.
97. Solaiman, D. K., Somkuti, G. A., and Steinberg, D. H. Construction and characterization of shuttle plasmids for lactic acid bacteria and *Escherichia coli*. *Plasmid* **28:** 25–36, 1992.
98. Somkuti, G. A., Solaiman, D. K., and Steinberg, D. H. Native promoter-plasmid vector system for heterologous cholesterol oxidase synthesis in *Streptococcus thermophilus*. *Plasmid* **33:** 7–14, 1995.
99. Sulavik, M. C., Tardif, G., and Clewell, D. B. Identification of a gene rgg which regulates expression of glucosyltransferase and influences the spp phenotype of *Streptococcus gordonii* Challis. *J. Bacteriol.* **174:** 3577–3586, 1992.
100. Suzuki, T., Shibata, C., Yamaguchi, A., Igarashi, K., and Kobayashi, H. Complementation of an *Enterococcus hirae (Streptococcus faecalis)* mutant in the α subunit of the H(+)-ATPase by cloned genes from the same and different species. *Mol. Microbiol.* **9:** 111–118, 1993.
101. Tannock, G. W., Luchansky, J. B., Miller, L., Connell, H., Thode-Andersen, S., Mercer A. A., and Klaenhammer, T. R. Molecular characterization of a plasmid-borne (pGT633) erythromycin resistance determinant (ermGT) from *Lactobacillus reuteri* 100–63. *Plasmid* **31:** 60–71, 1994.
102. van de Guchte, M., van der Vossen, J. M., Kok, J., and Venema, G. Construction of a lactococcal expression vector: expression of hen egg white lysozyme in *Lactococcus lactis* subsp. *lactis*. *Appl. Environ. Microbiol.* **55:** 224–228, 1989.
103. van der Vossen, J. M. B. M., Kok, J., and Venema, G. Construction of cloning promoter-screening and terminator-screening shuttle vectors for *Bacillus subtilis* and *Streptococcus lactis*. *Appl. Environ. Microbiol.* **50:** 540–542, 1985.
104. Vaughan, E. E., David, S., and de Vos, W. M. The lactose transporter in *Leuconostoc lactis* is a new member of the LacS subfamily of galactoside-pentose-hexuronide translocators. *Appl. Environ. Microbiol.* **62:** 1574–1582, 1996.
105. Vickerman, M. M., Heath, D. G., and Clewell, D. B. Construction of recombination-deficient strains of *Streptococcus gordonii* by disruption of the *recA* gene. *J. Bacteriol.* **175:** 6354–6357, 1993.
106. von Wright, A., Wessels, S., Tynkkynen, S., and Saarela, M. Isolation of a replication region of a large lactococcal plasmid and use in cloning of a nisin resistance determinant. *Appl. Environ. Microbiol.* **56:** 2029–2035, 1990.
107. Walker, D. C., and Klaenhammer, T. R. Isolation of a novel IS3 group insertion element and construction of an integration vector for *Lactobacillus* spp. *J. Bacteriol.* **176:** 5330–5340, 1994.
108. Watanabe, K., Hamasaki, M., Nakashima, Y., Kakita, Y., and Miake, F. High-frequency transformation of *Lactobacilllus casei* with plasmid pHY300PLK by electroporation. *Curr. Microbiol.* **29:** 217–222, 1994.
109. Wirth, R., An, F. Y., and Clewell, D. B. Highly efficient protoplast transformation sys-

tem for *Streptococcus faecalis* and a new *Escherichia coli-S. faecalis* shuttle vector. *J. Bacteriol.* **165:** 831–836, 1986.

110. Wyckoff, H. A., Barnes, M., Gillies, K. O., and Sandine, W. E. Characterization and sequence analysis of a stable cryptic plasmid from *Enterococcus faecium* 226 and development of a stable cloning vector. *Appl. Environ. Microbiol.* **62:** 1481–1486, 1996.
111. Yamamoto, N., and Takano, T. Isolation and characterization of a plasmid from *Lactobacillus helveticus* CP53. *Biosci. Biotechnol. Biochem.* **60:** 2069–2070, 1996.
112. Yuksel, G. U., and Steele, J. L. DNA sequence analysis expression distribution and physiological role of the Xaa-prolyldipeptidyl aminopeptidase gene from *Lactobacillus helveticus* CNRZ32. *Appl. Microbiol. Biotechnol.* **44:** 766–773, 1996.
113. Zink, A., Klein, J. R., and Plapp, R. Transformation of *Lactobacillus delbruckii* ssp. *lactis* by electroporation and cloning of origin of replicon by use of a positive selection vector. *FEMS Microbiol. Lett.* **78:** 207–212, 1991.

3.4 GENETIC RECOMBINATION IN LACTIC ACID BACTERIA

The following abbreviations are used in the data presented in this chapter:

amyA, α-amylase gene
cbh, conjugated bile acid hydrolase
celA, endoglucanase gene
cml, chloramphenicol-resistance gene
Cnj, conjugation
E, Enterococcus
Ep, electroporation
gftD, glucosyltransferase gene
int, integrase gene
IS, insertion sequence
lacM-lacL, β-galactosidase gene
lacZ, β-galactosidase gene
lacS, lactose permease gene
las operon, lactocin S operon
L, Lactobacillus
Lc, Lactococcus
Mob, mobilization
Pf, protoplast fusion
ptsH, HPr gene
ptsI, enzyme I gene
recA, recA protein gene
Rif^r, rifampicin resistance
scrA, sucrose-specific Ell permease gene
S, Streptococcus

St[r], streptomycin resistance
Tf, transformation
Tn, transposon

Insertion of Plasmid DNA by Homologous Recombination

Organism	Characteristics Acquired	Introduced Sequence Involved	Targeted Sequence in the Organism	Gene Transfer	Ref.
S. gordonii	Defective recombination ability	Conserved region of *recA* on a nonreplicative plasmid	Chromosomal *recA* gene	Ep	23
L. plantarum	Productivity of α-amylase and endoglucanase	Genomic DNA fragment on a nonreplicative plasmid with *amyA, celA*	Chromosomal DNA	Ep	18
L. sake	Defective PTS carbohydrates fermenting ability	*ptsI* gene fragment on a nonreplicative plasmid	Chromosomal *ptsI* gene	Ep	11
L. sake	Defective lactose fermenting ability	*lacL* gene fragment on a nonreplicative plasmid	Chromosomal *lacL* gene	Ep	11

Replacement of DNA Fragment by Homologous Recombination

Organism	Characteristics Acquired	Introduced Sequence Involved	Targeted Sequence in the Organism	Gene Transfer	Ref.
L. plantarum	Disrupted *cbh* gene	*cml*-containing *cbh* gene on a nonreplicative plasmid	Chromosomal *cbh* gene	Ep	10
S. mutans	Inactivated *gftD* gene	Interrupted *gft*D gene on a nonreplicative plasmid	Chromosomal *gftD* gene	Ep	8
S. sobrinus	Inactivated *scrA* gene	Interrupted *scrA* gene on an unstable plasmid	Chromosomal *scrA* gene	Mob	4
S. thermophilus	Defective lactose fermenting ability	Deleted *lacZ* gene on a nonreplicative plasmid	Chromosomal *lacZ* gene	Ep	15
S. thermophilus	Chloramphenicol resistance	Promoterless *cat* gene inserted between *lacS* and *lacZ* genes on a nonreplicative plasmid	Chromosomal *lacS-lacZ* genes	Ep	15
L. helveticus	Inactivated *pepXP* gene	Deleted *pepXP* gene on a temperature-sensitive plasmid	Chromosomal *pepXP* gene	Ep	2
L. acidophilus	β-galactosidase activity	Genomic DNA fragment having β-*gal* gene on a nonreplicative plasmid	Chromosomal DNA	Ep	12

Insertion of Transposable Element

Organism	Characteristics Acquired	Introduced Sequence Involved	Targeted Sequence in the Organism	Gene Transfer	Ref.
L. casei	Conversion of a temperate phage into a lytic phage	Endogenous ISL*1*	Temperate phage genome on chromosome		19
L. helveticus	Defective lactose fermenting ability	Endogenous ISL*2*	*lacL-lacM* locus on chromosome		26
S. gordonii	Defective coaggregating ability	Tn*916* carried on a plasmid	Chromosomal DNA	Tf	25
S. gordonii	Gentamycin resistance	Tn*4001* on a non-replicative plasmid	Chromosomal DNA	Ep	13
L. sake	Abolished bacteriocin production	Endogenous IS*1163*	*las* operon on resident plasmid		21
L. plantarum	Erythromycin resistance	Tn*917* on a temperature-sensitive plasmid	Resident plasmid	Ep	7
E. faecalis *S. sanguis* *S. pneumoniae* *S. pyogenes*	Erythromycin resistance	Conjugative chromosomal element Ω6001 bearing *erm* gene	Chromosomal DNA	Cnj	16
S. *crista*	Reduced ability of binding to cells of Fusobacterium nucleatum	Tn*916* on a plasmid	Chromosomal DNA	Tf	6

Insertion of Plasmid DNA via Transposition of Transposable Element

Organism	Characteristics Acquired	Introduced Sequence Involved	Targeted Sequence in the Organism	Gene Transfer	Ref.
E. faecalis *S. thermophilus*	Erythromycin resistance	ISS*1* on a temperature sensitive plasmid	Chromosomal DNA	Ep	14
L. gasseri	Erythromycin resistance	IS*1223* on a nonreplicative plasmid having *erm* gene	Chromosomal DNA	Ep	24
E. faecalis	Erythromycin resistance	*att* site of Tn*1545* on a non-replicative plasmid with a provision of integrase in trans	Chromosomal DNA	Ep Cnj	22

(continued)

Insertion of Plasmid DNA by Phage Function

Organism	Characteristics Acquired	Introduced Sequence Involved	Targeted Sequence in the Organism	Gene Transfer	Ref.
E. faecalis *L. casei* *L. lactis* *S. pneumoniae*	Erythromycin resistance	*attP* site of phage mv4 on a nonreplicative plasmid with *int* gene	Chromosomal DNA	Ep	1
L. gasseri	Erythromycin resistance	*attP* site of phage ϕadh on a nonreplicative plasmid with *int* gene	Chromosomal DNA	Ep	17
S. thermophilus	Chloramphenicol resistance	*attP* site of phage ϕSfi21 on a nonreplicative plasmid with *int* gene	Chromosomal DNA	Ep	3

Unidentified Mechanism

Organism	Characteristics Acquired	Introduced Sequence Involved	Targeted Sequence in the Organism	Gene Transfer	Ref.
L. fermentum	Trehalose fermenting ability	Chromosomal DNA of trehalose-positive *Lc. lactis* harboring plasmid pAMβ1	Chromosomal DNA	Pf	5
L. casei	Double resistance to both rifampicin and streptomycin	Chromosomal DNA of Rif^r strain	Chromosomal DNA of St^r strain	Pf	9
E. faecalis *L. plantarum*	Double resistance to both chloramphenicol and erythromycin	Cointegrant with conjugative plasmid pAMβ1 formed in *E. faecalis*	Successful or unsuccessful resolution of cointegrant in *L. plantarum*	Mob	20

REFERENCES

1. Auvray, F., Coddeville, M., Ritzenthaler, P., and Dupont, L. Plasmid integration in a wide range of bacteria mediated by the integrase of *Lactobacillus delbrueckii* bacteriophage mv4. *J. Bacteriol.* **179:** 1837–1845, 1997.
2. Bhowmik, T., Fernandez, L., and Steele, J. L. Gene replacement in *Lactobacillus helveticus*. *J. Bacteriol.* **175:** 6341–6344, 1993.
3. Bruttin, A., Foley, S., and Brussow, H. The site-specific integration system of the temperate *Streptococcus thermophilus* bacteriophage ϕSfi21. *Virology* **237:** 148–158, 1997.
4. Buckley, N. D., Lee, L. N., and LeBlanc, D. J. Use of a novel mobilizable vector to inactivate the *scrA* gene of *Streptococcus sobrinus* by allelic replacement. *J. Bacteriol.* **177:** 5028–5034, 1995.

5. Cocconcelli, P. S., Morelli, L., Vescovo, M., and Bottazzi, V. Intergeneric protoplast fusion in lactic acid bacteria. *FEMS Microbiol. Lett.* **35:** 211–214, 1986.
6. Correia, F. F., DiRienzo, J. M., Lamont, R. J., Anderman, C., McKay, T. L., and Rosan, B. Insertional inactivation of binding determinants of *Streptococcus crista* CC5A using Tn916. *Oral Microbiol. Immunol.* **10:** 220–226, 1995.
7. Cosby, W. M., Axelsson, L. T., and Dobrogosz, W. J. Tn*917* transposition in *Lactobacillus plantarum* using the highly temperature-sensitive plasmid pTV1Ts as a vector. *Plasmid* **22:** 236–243, 1989.
8. Hanada, N., and Kuramitsu, H. K. Isolation and characterization of the *Streptococcus mutans gtfD* gene, coding for primer-dependent soluble glucan synthesis. *Infect. Immun.* **57:** 2079–2085, 1989.
9. Kang, Y., Kim, J. H., and Ryu, D. D. Y. Protoplast fusion of *Lactobacillus casei. Agric. Biol. Chem.* **51:** 2221–2227, 1987.
10. Leer, R. J., Christiaens, H., Verstraete, W., Peters, L., Posno, M., and Pouwels, P. H. Gene disruption in *Lactobacillus plantarum* strain 80 by site-specific recombination: Isolation of a mutant strain deficient in conjugated bile salt hydrolase activity. *Mol. Gen. Genet.* **239:** 269–272, 1993.
11. Leloup, L., Ehrlich, S. D., Zagorec, M., and Morel-Deville, F. Single-crossover integration in the *Lactobacillus sake* chromosome and insertional inactivation of the *ptsI* and *lacL* genes. *Appl. Environ. Microbiol.* **63:** 2117–2123, 1997.
12. Lin, M. Y., Harlander, S., and Savaiano, D. Construction of an integrative food-grade cloning vector for *Lactobacillus acidophilus. Appl. Microbiol. Biotechnol.* **45:** 484–489, 1996.
13. Lunsford, R. D. A Tn*4001* delivery system for *Streptococcus gordonii* (Challis). *Plasmid* **33:** 153–157, 1995.
14. Maguin, E., Prevost, H., Ehrlich, S. D., and Gruss, A. Efficient insertional mutagenesis in lactococci and other gram-positive bacteria. *J. Bacteriol.* **178:** 931–935, 1996.
15. Mollet, B., Knol, J., Poolman, B., Marciset, O., and Delley, M. Directed genomic integration, gene replacement, and integrative gene expression in *Streptococcus thermophilus. J. Bacteriol.* **175:** 4315–4324, 1993.
16. Pozzi, G., Musmanno, R. A., Renzoni, E. A., Oggioni, M. R., and Cusi, M. G. Host-vector system for integration of recombinant DNA into chromosomes of transformable and nontransformable streptococci. *J. Bacteriol.* **170:** 1969–1972, 1988.
17. Raya, R. R., Fremaux, C., de Antoni, G. L., and Klaenhammer, T. R. Site-specific integration of the temperate bacteriophage ϕadh into the *Lactobacillus gasseri* chromosome and molecular characterization of the phage (attP) and bacterial (attB) attachment sites. *J. Bacteriol.* **174:** 5584–5592, 1992.
18. Scheirlinck, T., Mahillon, J., Joos, H., Dhaese, P., and Michiels, F. Integration and expression of alpha-amylase and endoglucanase genes in the *Lactobacillus plantarum* chromosome. *Appl. Environ. Microbiol.* **55:** 2130–2137, 1989.
19. Shimizu-Kadota, M., Kiwaki, M., Hirokawa, H., and Tsuchida, N. ISL*1*: A new transposable element in *Lactobacillus casei. Mol. Gen. Genet.* **200:** 193–198, 1985.
20. Shrago, A. W., and Dobrogosz, W. J. Conjugal transfer of group B streptococcal plasmids and comobilization of *Escherichia coli-Streptococcus* shuttle plasmids to *Lactobacillus plantarum. Appl. Environ. Microbiol.* **54:** 824–826, 1988.
21. Skaugen, M., and Nes, I. F. Transposition in *Lactobacillus sake* and its abolition of lac-

tocin S production by insertion of IS*1163*, a new member of the IS*3* family. *Appl. Environ. Microbiol.* **60:** 2818–2825, 1994.

22. Trieu-Cuot, P., Carlier, C., Poyart-Salmeron, C., and Courvalin, P. An integrative vector exploiting the transposition properties of Tn*1545* for insertional mutagenesis and cloning of genes from gram-positive bacteria. *Gene* **106:** 21–27, 1991.
23. Vickerman, M. M., Heath, D. G., and Clewell, D. B. Construction of recombination-deficient strains of *Streptococcus gordonii* by disruption of the *recA* gene. *J. Bacteriol.* **175:** 6354–6357, 1993.
24. Walker, D. C., and Klaenhammer, T. R. Isolation of a novel IS3 group insertion element and construction of an integration vector for *Lactobacillus* spp. *J. Bacteriol.* **176:** 5330–5340, 1994.
25. Whittaker, C. J., Clemans, D. L., and Kolenbrander, P. E. Insertional inactivation of an intrageneric coaggregation-relevant adhesin locus from *Streptococcus gordonii* DL1 (Challis). *Infect. Immun.* **64:** 4137–4142, 1996.
26. Zwahlen, M. C., and Mollet, B. ISL*2*, a new mobile genetic element in *Lactobacillus helveticus. Mol. Gen. Genet.* **245:** 334–338, 1994.

4

ROLE OF PROBIOTICS IN HEALTH AND DISEASES

4.1 HUMAN AND ANIMAL MODELS

4.1.1 Prevention and Treatment of Gastrointestinal Bacterial Infection and Diseases

Effect	Subject	Probiotics	Route of Administration[a]	Viability	Ref.
Protect against *Salmonella typhimurium* infection	Conventional mice	Indigenous bacteria	Intravenously	Viable	17
	Male Sprague-Dawley rat	Yogurt bacteria	Oral	Viable	7
Protect against *Listeria monocytogenes* infection	Male BALB/c mice	*Lactobacillus casei* strain Shirota	ip	Polysaccharide peptidoglycan extract	13
Protect against *Pseudomonas aeroginosa, Escherichia coli, Salmonella typhimurium* infection	Male BALB/c mice	*L. casei* strain Shirota	ip	Heat-killed	13

(*continued*)

Effect	Subject	Probiotics	Route of Administration[a]	Viability	Ref.
Treatment of diarrheal disorders in children	Children with persistent diarrhea	Yogurt	Oral	Viable	4
		L. rhamnosus GG	Oral	Viable	9, 16, 11
		Streptococcus faecium	Oral	Viable	1
		Bifidobacterium breve	Oral	Viable	8
No positive clinical benefit	Premature infants	*L. rhamnosus* GG	Oral	Viable	12
Prevention of traveler's diarrhea	Healthy adult	*L. rhamnosus* GG	Oral	Viable	15
Prevention of antibiotic associated diarrhea	Elderly outpatients	Lactinex: *L. acidophilus* + *L. bulgaricus*	Oral / iv	Viable	6
	Elderly hospital inpatients	*S. faecium*	Oral	Viable	3
Treatment of relapsing *Clostridium difficile* colitis	Adult outpatients	*L. rhamnosus* GG	Oral	Viable	5
	Children outpatients	*L. rhamnosus* GG	Oral	Viable	2
Preservation of intestinal integrity during radiotherapy	Adult outpatients	*L. acidophilus*	Oral	Viable	18
Protection against lethal irradiation	Male C3H/HeN mice	*L. casei* strain Shirota	sc	Heat-killed	14, 19
Prevention of *Helicobacter pylori* infection	Male BALB/c mice	*L. salivarius* WB 1004	Oral	Viable	10

[a]ip, intraperitoneal; iv, intravenous; sc, subcutaneous.

Protection Against Salmonella *Infection*

Effect on Number of Salmonella *in Spleen after Intravenous Challenge (17)*

Indigenous Microbes Isolated from Mice and Culture Media Used

Bacteria	Characteristics	Culture[a]
Lactobacillus 100-5	Colonizes nonsecretory epithelium of stomach of monoassociated CD-1 and BALB/c mice biotype A	PRAS BHI + 0.1% Tween 80, 48h
Lactobacillus 100-20	*L. leichmannii* biotype D	PRAS BHI + 0.1% Tween 80, 48h

(*continued*)

Indigenous Microbes Isolated from Mice and Culture Media Used (*cont.*)

Bacteria	Characteristics	Culture[a]
Bacteroides 116-6	Some resemblance to *B. fragilis* biotype A	PRAS BHI, 48h
Bacteroides 116-16	Some resemblance to *B. fragilis* biotype G	PRAS BHI, 48h
Fusiform 999-1	Some resemblance to *Fusobacterium russii*	PRAS SE, 72h
Fusiform 999-12	*Clostridium* sp. biotype D	PRAS SE, 72h
Fusiform 999-17	*Clostridium* sp. biotype F	PRAS SE, 72h
Spirals 120-1 and 120-3	Isolated from mucosal scrapings of cecum, some resemblance to *Campylobacter faecalis*	EM (no blood), 7 days
Coliform 104-1	*E. coli* biotype E	BHI, 48h
Coliform 104-2	*E. coli* biotype D	BHI, 48h
Enterococcus 102-11	*S. faecalis* biotype A	BHI, 48h
Enterococcus 102-12	*S. faecium.* Xylose fermenting biotype C	BHI, 48h

[a]PRAS, prereduced anaerobically sterilized; BHI, brain–heart infusion; SE, sweet E; EM, E medium.

Number of *Salmonella* in Spleen of Conventional Mice 5 Days after Intravenous Challenge

Inoculum[a]	No. of Experiments	Total No. of Mice	Mean $\log_{10}$ *Salmonella* per Spleen ± SD
BCG	6	30	3.1 ± 0.3[b]
100-20	4	20	4.6 ± 0.3[b]
102-12	6	30	4.6 ± 0.3[b]
104-1	4	20	4.7 ± 0.3[b]
102-11	4	20	4.7 ± 0.4[b]
116-6	4	20	4.8 ± 0.3[b]
104-2	4	20	4.8 ± 0.2[b]
SE	3	15	5.0 ± 0.2
999-17	1	5	5.0 ± 0.1
999-1	1	5	5.1 ± 0.3
120-1	3	15	5.2 ± 0.3
100-5	3	15	5.2 ± 0.3
120-3	3	15	5.3 ± 0.4
999-12	1	5	5.4 ± 0.4
EM	3	15	5.4 ± 0.4
BHI (anaerobic +0.1% Tween 80)	3	15	5.5 ± 0.5
BHI (aerobic)	3	15	5.5 ± 0.5
Control (no inoculum)	11	55	5.5 ± 0.4
BHI (anaerobic)	3	15	5.5 ± 0.6
116-16	1	5	5.9 ± 0.7

[a]Male, conventional mice (7–8 weeks old) were inoculated intravenously with 0.1 ml of 1×10^6 viable organisms of the indigenous bacterial strain or 1×10^6 BCG mycobacteria. The mice were then intravenously challenged with 5×10^3 *S. typhimurium* LT2 3 weeks later. The mice were killed 5 days after challenge with *S. typhimurium,* spleen removed, weighed, and homogenized. Aliquots of dilutions were cultured as pour plates in brain–heart infusion (Difco) agar.
[b]Significantly different from control (Student's *t*-test, 1% level).

Remarks Inoculation of conventional mice with viable indigenous bacterial strains or BCG vaccine resulted in a lower number of *Salmonella* in the spleen following challenge compared with control mice. This shows that some types of indigenous bacteria can influence host resistance to *Salmonella* infections in systemic tissues. Since the indigenous bacteria do not persist long enough in the spleen following intravenous inoculation for there to be a direct antagonistic effect between the indigenous bacteria and the pathogen, it is believed that the indigenous microbes influence the immunological mechanisms of the host.

Effect on Mortalities for Rats Ingesting Salmonella *Orally (7)*

	Time Postinoculation (days)[a]			
	1	7	14	21
Pooled cumulative mortalities				
For [b]M-rats (n = 104)	1	4	20	22
Y-rats (n = 107)	1	2	5	7

[a]*Salmonella enteritidis* serotype *typhimurium,* obtained from R.W. Wannemacher, Jr., (USAMIRID, Ft. Detrick, Frederick, MD) and was subcultured in brain–heart infusion broth at 37°C. Rats were anaesthetized with carbon dioxide and inoculated intragastrically per os with a saline suspension of the pathogen.
[b]Male weaning Sprague-Dawley outbred rats (initial weight about 50 g) were obtained from Dominion Labs. (Dublin, VA). The diets consisted of freeze-dried fortified milk (M-rats) or the corresponding freeze-dried yogurt made from the control milk (Y-rats). The milk was standardized to 1.5% fat and fortified with 3% nonfat dry milk, iron (85 ppm) and copper (8 ppm) salts and vitamin mixture (20% of the manufacturer's recommended level).

Remarks Mean death time postchallenge was about 10 days for rats fed either diet, but yogurt-fed (Y) rats were clearly more resistant than milk-fed (M) rats. Two-thirds of the mortalities of M-rats occurred in week 2 postchallenge, while the Y-rats mortalities tended to be later, that is, in week 3. Thus the cumulative rate and extent of death were greater for M-rats than for Y-rats.

Protection Against* Listeria *Infection (13)

Day of Treatment before Challenge[a]	Treatment[a]	Log_{10} No. of Viable Bacteria in the Peritoneal Cavity (mean ± SD)[b]	Survival Rate (%)[c]
—	None (control)	5.71 ± 0.16	0/5 (0)
–3	PSPG	4.83 ± 0.06[d]	5/5 (100)
	LC 9018	5.32 ± 0.11[d]	4/5 (80)
–5	PSPG	4.49 ± 0.08[d]	5/5 (100)
	LC 9018	4.57 ± 0.14[d]	5/5 (100)
–7	PSPG	4.23 ± 0.28[d]	5/5 (100)
	LC 9018	4.14 ± 0.08[d]	5/5 (100)
–10	PSPG	4.62 ± 0.06[d]	5/5 (100)
	LC 9018	4.04 ± 0.14[d]	5/5 (100)

[a]Polysaccharide–peptidoglycan complex extracted from *L. casei* strain Shirota (PSPG) or LC 9018 (*L. casei* strain Shirota heat-killed) at a dose of 0.5 mg per mouse was injected ip into male BALB/c mice (7

weeks old) purchased from Shizuoka Agricultural Cooperative for Experimental Animals Japan. *L. casei* strain Shirota was provided by Yakult Central Institute for Microbiology Research, Tokyo.
[b]*L. monocytogenes* EGD [5.0×10^6 colony-forming units (CFU) per mouse] was injected ip into 10 mice of each group. Five mice per group were dissected 24 h later, and
[c]The remaining 5 mice were observed for survival for 14 days after the challenge injection.
[d]Difference from the untreated control was significant; $p < 0.01$.

Remarks Treatment of mice with PSPG 3 to 10 days earlier augmented the resistance of the mice to a lethal infection with *L. monocytogenes* (5.0 × 106 CFU). The most effective resistance was shown in the mice when they were treated with PSPG for 5–7 days before infection. The antilisterial effect of PSPG was detected earlier after administration than that of LC 9018.

Protection Against Other Gram-Negative Bacteria Infection

Effect on P. aeruginosa *Infection (13)*

Day of Treatment before Challenge[a]	Log_{10} No. of Viable Bacteria in Organ (mean ± SD)[b]		Survival Rate (%)[c]
	Peritoneal Cavity	Liver	
Untreated	7.24 ± 1.04 (6/6)[d]	6.17 ± 0.73 (6/6)	2/10 (20)
−1	3.70 ± 1.35[e] (6/6)	3.33 (1/6)	10/10 (100)
−3	3.02 (2/6)	< 2.3 (0/6)	10/10 (100)
−5	3.86 ± 0.72[e] (3/6)	4.54 (2/6)	10/10 (100)
−7	3.25 (2/6)	3.41 (2/6)	10/10 (100)
−10	4.03 ± 0.33[e] (3/6)	4.09 ± 0.86[e] (5/6)	10/10 (100)

[a]PSPG at a dose of 1 mg per mouse was injected ip into male BALB/c mice.
[b]*P. aeruginosa* KC-2 (3.7×10^6 CFU per mouse) was injected ip into 16 mice in each group. Six mice per group were dissected 18 h later.
[c]The remaining 10 mice were observed for survival for 14 days after the challenge injection.
[d]Numbers in parentheses represent number of mice in which bacteria were determined divided by number of mice dissected.
[e]Difference from the untreated control was significant; $p < 0.01$.

Remarks Treatment of mice with PSPG for 1–10 days earlier enhanced the resistance of the mice to a lethal infection with *P. aeruginosa* (3.7×10^6 CFU) and the growth of the bacteria in the peritoneal cavity and liver was inhibited markedly in PSPG-treated mice.

Effect on P. aeruginosa, E. coli *and* S. typhimurium *Infection (13)*

Treatment[a]	Log_{10} No. of Viable Bacteria in Organ (mean ± SD)[b]	
	Peritoneal Cavity	Liver
P. aeruginosa		
None (control)	8.28 ± 0.61	7.69 ± 0.20
PSPG	3.08 ± 0.48[c]	< 2.3
LC 9018	5.46 ± 0.67[c]	4.37 ± 1.01[c]

(*continued*)

Treatment[a]	Log_{10} No. of Viable Bacteria in Organ (mean ± SD)[b]	
	Peritoneal Cavity	Liver
E. coli		
None (control)	4.78 ± 0.57	6.77 ± 0.38
PSPG	3.71 ± 0.76	5.01 ± 0.16[c]
S. typhimurium		
None (control)	6.18 ± 0.13	NT[d]
PSPG	4.75 ± 0.19[c]	NT[d]
LC 9018	4.23 ± 0.17[c]	NT[d]

[a]PSPG (1 mg per mouse) or LC 9018 (0.5 mg per mouse) was injected ip into male mice 5 days before the challenge with bacteria. To prepare PSPG from *L. casei,* lyophilized preparation of heat-killed (100°C for 30 min) *L. casei* strain Shirota (LC 9018) was used. One gram of LC 9018 was suspended in 100 ml of Tris-maleate buffer (pH 6.4) supplemented with 5 mM $MgCl_2$, and then 10 mg of *N*-acetylmuramidase was added. After 4 h at 37°C, the suspension was centrifuged at 10,000 g for 15 min and the supernatant was incubated at 37°C for another 16 h for complete digestion. The suspension was then treated with 10 mg of DNase and RNase at 37°C for 16 h and finally incubated with 10 mg of trypsin at 37°C for 16 h. The mixture was then dialyzed against distilled water at 4°C for 2 days, and the inner solution was lyophilized and used as PSPG.
[b]Mice were dissected 24 h after the ip challenge with bacteria. Inoculum: *P. aeruginosa* KC-2, 1.3×10^7; *E. coli* E77156 06, 5.0×10^6; *S. typhimurium* LT-2, 3.4×10^5 (CFU per mouse). To test for bacterial growth, the peritoneal cavity was washed with 5 ml of saline; the peritoneal cells were lysed by repeated freezing and thawing, and the fluid was diluted 10-fold with saline. Each dilution (0.1 ml) was spread on to a nutrient agar plate containing 0.4% (w/v) glucose. The number of colonies in 5 ml of harvested fluid was expressed as $\log_{10}$ CFU. Bacterial growth in the liver was determined by spreading organ homogenates with a conventional Teflon homogenizer.
[c]Difference from the untreated control was significant; $p < 0.01$.
[d]Not tested.

Remarks Mice that had received 1 mg of PSPG or LC 9018 ip 5 days earlier were injected ip with *P. aeruginosa* (1.3×10^7 CFU), *E. coli* (5.0×10^6 CFU), or *S. typhimurium* (3.4×10^5 CFU), and the numbers of viable bacteria in the peritoneal cavity and liver of the infected mice were determined. PSPG augmented the resistance of mice to infections with Gram-negative bacteria at an early phase of the infection.

Treatment of Diarrheal Disorders in Children

Effect of Yogurt (4)

Clinical Characteristics of Children Included in Investigation

	Total	Milk	Yogurt
Randomized subjects	52	27	25
Excluded subjects	7 (13%)	3 (11%)	4 (16%)
Sex (M/F)	32/20	17/10	17/8
Age (months)	7.2 ± 4.6	6.9 ± 4.5[a]	7.6 ± 4.9[a]
Weight (kg)	6.4 ± 2.3	6.2 ± 2.4[a]	6.8 ± 2.1[a]
Weight/height (% NCHS[b])	90 ± 11	89 ± 10[a]	91 ± 13[a]

(continued)

Clinical Characteristics of Children Included in Investigation (*cont.*)

	Total	Milk	Yogurt
Duration of diarrhea before admission (days)	20 ± 6	20 ± 5[a]	21 ± 6[a]
No. of vomiting episodes in the last 24 h	0.7 ± 1.0	0.8 ± 1.2[a]	0.6 ± 1.0[a]
No. of stools in the last 24 h	5.3 ± 2.0	5.9 ± 2.2[a]	4.6 ± 1.4[a]

[a]No statistically significant differences were found between the two groups.
[b]NCHS: National Centre for Health Statistics.
Reproduced with permission of Lippincott Williams & Wilkins.

Experimental Protocol. The milk used was the formula available in Algeria. Yogurt was prepared in the clinical unit, using the same formula, by fermentation with a specific freeze-dried starter culture (*S. thermophilus* and *L. bulgaricus*) and stored at 8°C for the 5-day study period. In addition, both groups received cereals diluted in water (12% w/v) with added sucrose (5% w/v) and vegetable (50% w/v) with fat (12% w/v) added in the form of butter. Each child was given a total for all foods of 180 ml/kg/day in 6 meals to ensure a total caloric intake of 150–180 kcal/kg/day. The repartition of meals varied with age as follows: between 3 and 6 months of age, the number of servings of milk or yogurt, vegetables, and cereals was 4, 1, and 1; between 6 and 12 months, 3, 2, and 1 servings; between 12 and 36 months, 2, 2, and 2 servings, respectively. Each stool was collected in diapers (urine collected separately) and immediately weighed. Cessation of diarrhea was declared after the last liquid or semiliquid stool evacuated before two formed stools.

Number of Children with Persistent Diarrhea Who Presented with Clinical Failure after Milk or Yogurt Feedings

	Milk ($n = 24$)	Yogurt ($n = 21$)
Duration of diarrhea > 5 days	8	3
Weight loss > 5% during 24 h	2	0
Total number of clinical failures	10	3 ($p < 0.05$)

Reproduced with permission of Lippincott Williams & Wilkins.

Clinical Outcome of 45 Children with Persistent Diarrhea after Milk or Yogurt Feedings[a]

	Milk ($n = 24$)	Yogurt ($n = 21$)
Total liquid stool output (g)	1000 (359–1641)	853 (479–1227)
Total ORS[b] intake during diarrhea (ml)	1112 (575–1649)	445 (251–639)
Total energy intake during diarrhea (kcal)	1850 (1254–2446)	1430 (883–1977)
Gain in weight after 5 days (% initial weight)	1.2 (–0.4–2.8)[c]	2.1 (0.3–3.9)

Reproduced with permission of Lippincott Williams & Wilkins.
[a]Results are shown as the mean (95% confidence limits).
[b]ORS, oral rehydration solution.
[c]$n = 22$, two children with a weight loss > 5% during 24 h were excluded.

Remarks Clinical failure was observed in 10 of the children receiving milk (42%) and only 3 of those receiving yogurt (14%). None of the initial clinical characteristics of those children fed successfully versus those fed unsuccessfully were significantly different. The beneficial effect of yogurt was further substantiated by the analysis of the other outcome variables, which showed a tendency of a greater weight gain despite a lower energy input, a smaller liquid stool output, and a smaller ORS intake in yogurt-fed children versus milk-fed children. The benefit can be mainly explained by: (1) 44% decrease in lactose content of yogurt compared to milk, and the more efficient lactose digestion from yogurt than from milk (due to intraluminal activity of the microbial β-galactosidase contained in yogurt). (2) Cow's milk proteins are modified in yogurt. (3) Yogurt stimulates immunity. (4) Yogurt may have probiotic properties. Furthermore, its low pH and the large quantity of viable bacteria it contains make yogurt a food product free from pathogenic bacterial contamination. This latter property may help prevent children consuming yogurt from becoming reinfected, which is probably an important factor in persistent diarrhea.

Effect of L. rhamnosus

Experimental Protocol (9)

1. Seventy-one well-nourished patients (56% males) between 4 and 45 months of age consecutively admitted at Tampere University (n = 56) or Satakunta Central University, Finland (n = 15) for acute diarrhea of less than 7 days' duration and with more than 3 watery stools during the previous 24 hours were the study subjects. The study was carried out between January and May 1989. Eight control patients between 2 and 17 months were also examined.
2. Acute weight loss was calculated as the difference between expected weight (according to individual growth charts) and observed weight. Fluid deficit (dehydration percent) was then defined from the clinical signs of dehydration and acute weight loss with a reduction of 0.5–1% per day if the diarrhea had continued for at least 3 days to reflect loss of weight due to low caloric intake. Feces were tested for rotavirus antigen using an enzyme immunoasssay. Oral rehydration was accomplished in 6 hours.
3. The children were randomly allocated to receive one of the three dietary treatments. Group 1 (n = 24) was given *L. rhamnosus* strain GG-fermented milk product (lactose was hydrolyzed), 125 g (10^{10-11} CFU) twice daily. *Lactobacillus* GG was given to group 2 (n = 23) as a freeze-dried powder; one dose (10^{10-11} CFU) twice daily. Group 3 (n = 24) was given a placebo (fermented-then-pasteurized) yogurt with an insignificant amount of lactic acid bacteria, 125 g twice daily; each diet was given for 5 days.
4. In addition to the milk products, children younger then 1 year received jelly, mashed vegetable, and potato with meats as well as cooked cereal. The older patients received an ordinary mixed diet. The infants received their usual drink with their meals. If diarrhea continued, the losses were replaced by the rehydration solution together with free water.

Clinical Characteristics and Outcome of Therapy on Patients Receiving Rapid Realimentation with *Lactobacillus* GG-Fermented Milk Product (Group 1), *Lactobacillus* GG Powder (Group 2), or a Placebo Milk Product with No Lactic Acid Bacteria (Group 3)[a]

	Group 1 ($n = 24$)		Group 2 ($n = 23$)		Group 3 ($n = 24$)		Analysis of Variance	
Characteristic	Mean	SD	Mean	SD	Mean	SD	F	p
Age (months)	17.7	8.6	10.9	5.9	16.9	1.3	6.43	0.003
Duration of diarrhea at home (days)	2.5	1.2	2.4	0.9	2.7	1.0	0.60	0.55
Acute weight loss (g)	502	134	426	219	513	148	1.79	0.17
Dehydration (%)	4.6	1.2	4.7	2.4	4.9	1.3	0.2	0.82
Outcome								
Oral rehydration, solution given, ml/6 h	892	234	741	346	880	218	2.22	0.12
Weight gain during rehydration, g	242	172	136	236	273	302	2.05	0.14
Weight gain during realimentation, g	302	287	129	256	262	326	2.27	0.11
Duration of diarrhea, in hospital, days	1.4	0.8	1.4	0.8	2.4	1.1	8.70	<0.001

Reproduced by permission of *Pediatrics.*
[a]All obtained from the Valio Finnish Co-operative Dairies' Association, Findland.

Percentage of Patients with Watery Diarrhea

Days of Dietary Therapy	Group 1 ($n = 24$)	Group 2 ($n = 23$)	Group 3 ($n = 24$)	χ^2 Test	p
Day 1	96	96	96		
Day 2	50	52	79	5.28	0.07
Day 3	12.5	22	58	13.15	0.001
Day 4	0	4	21	7.47	0.02
Day 5	0	0	4	1.99	0.37

Percentage of Patients with Vomiting

Days of Dietary Therapy	Group 1 ($n = 24$)	Group 2 ($n = 23$)	Group 3 ($n = 24$)	χ^2 Test	p
Day 1	58	43	54	1.1	0.58
Day 2	21	22	38	2.14	0.34
Day 3	0	9	17	4.31	0.12
Day 4	0	0	0		
Day 5	0	0	0		

Remarks *Lactobacillus* GG administered either as a fermented milk product or as a freeze-dried powder reduced the duration of diarrhea by mean [95% confidence intervals (CI)] of 1.0 (0.4, 1.6) day. The effect of *Lactobacillus* GG on watery diarrhea became apparent after the first day of treatment. There was no more vomiting in the patients receiving *Lactobacillus* GG (groups 1 and 2) than in patients receiving the placebo milk product with no lactic acid bacteria. Thus *Lactobacillus* GG-fermented milk product with lactic acid bacteria that are able to colonize the gut is beneficial for the recovery from diarrhea.

Experimental Protocol (16)

1. The study was conducted in the diarrhea ward of the Infectious Diseases Hospital in Peshawar, Pakistan, between July and August 1993, the warm rainy season when incidence of acute infectious diarrhea is usually peak.
2. Forty children (25 boys and 15 girls) between 1 and 24 months old admitted with acute diarrhea were randomized to receive either *Lactobacillus* GG (supplied by Scientific Hospital Supplies, Liverpool, UK.), 10^{10-11} CFU as freeze-dried preparation or a placebo (microcrystalline cellulose) of identical appearance that was mixed in 10 ml of ORS and given orally, in addition to the usual diet after the initial rehydration period of 4–6 h. Treatment continued at 12 hourly intervals for 2 days.

Comparison of Characteristics and Clinical Response of Patients

Characteristic	Lactobacillus Group (n = 21)	Placebo Group (n = 19)	p
Age (mo)	11.5 ± 6.8[a]	14.4 ± 7.0	NS[b]
Weight at admission (kg)	6.9 ± 1.9	6.8 ± 2.3	NS
Duration of diarrhea preceding to admission (days)	4.5 ± 1.8	3.8 ± 2.1	NS
Stool frequency			
Day 1	8.5 ± 4.5	10.5 ± 5.1	NS
Day 2	5.8 ± 3.1	7.0 ± 3.3	NS
Frequency of vomiting			
Day 1	5.6 ± 2.4	5.8 ± 3.2	NS
Day 2	2.0 ± 1.5	4.0 ± 2.5	< 0.05
Weight gain (%) change after rehydration	7.1 ± 4.1	7.0 ± 3.5	NS
24 h posthydration	–0.6 ± 4.8	–1.4 ± 3.9	NS
48 h posthydration	+1.0 ± 5.0	–0.9 ± 4.7	NS
Stool output (boys only) (g/kg)	(n = 15)	(n = 10)	
Day 1	102.2 ± 45.6	88.6 ± 36.0	NS
Day 2	66.5 ± 31.3	54.1 ± 26.2	NS
	(n = 16)	(n = 16)	
No. of patients with persistent watery stool at 48 h	5 (31)[c]	12 (75)[c]	<0.01
Stool frequency on day 2	4.4 ± 2.0	6.6 ± 4.2	<0.05

[a]Mean ± SD.
[b]NS, not significant.
[c]Numbers in parentheses, percent.

Remarks There was less frequent vomiting on the second day in the *Lactobacillus* group. When those presenting with watery, nonbloody diarrhea were considered ($n = 32$), there was a significant reduction in the stool frequency and in the number of patients with persistent watery stool on the second day in the group receiving *Lactobacillus* GG. This suggests that the use of *Lactobacillus* GG may accelerate recovery in cases of acute watery diarrhea in children.

Experimental Protocol (11)

1. Forty-nine children (25 boys and 24 girls) between 4 and 35 months of age admitted for acute gastroenteritis of < 7 days' duration at the Department of Pediatrics, Tampere University Hospital, Finland, were enrolled in this study between February and June 1993, during a rotavirus epidemic.
2. On admission, they were weighed and examined clinically. The degree of dehydration (%) was determined from acute weight loss and clinical signs. The patients were orally rehydrated within 6 h. The children were then randomly and double-blindly allocated to receive one of the three lactic acid bacteria preparations twice daily for 5 days. The first group ($n = 16$) was given freeze-dried LGG, supplied by Valio ltd., Helsinki, Finland; the second group ($n = 14$) was given *L. casei* subsp. *rhamnosus* (Lactophilus), produced by Laboratoires Lyocentre, France; and the third group ($n = 19$) was given a combination of 95% *S. thermophilus,* 4% *L. delbrueckii* subsp. *bulgaricus*, and 1% *L. casei* subsp. *rhamnosus* (Yalacta), purchased from Yalacta, Caen Cedex, France.

Clinical Characteristics and Responses of Patients[a]

Characteristic	LGG ($n = 16$)	Lactophilus ($n = 14$)	Yalacta ($n = 19$)	p
Age (mo)	21.3 (9.5)	19.4 (8.4)	16.6 (9.2)	0.31[b]
Duration of symptoms at home	2.2 (1.7)	2.3 (2.3)	2.5 (1.7)	0.89[b]
Dehydration (%)	4.8 (2.2)	3.8 (0.8)	4.4 (1.5)	0.38[b]
Duration of diarrhea (days)	1.8 (0.8)	2.8 (1.2)	2.6 (1.4)	0.04[b]
Watery diarrhea				
Day 1	11 (69)	13 (93)	13 (68)	0.20[c]
Day 2	3 (19)	9 (64)	11 (58)	0.02[c]
Day 3	0 (0)	3 (21)	6 (32)	0.05[c]
Vomiting				
Day 1	10	5	9	0.34[c]
Day 2	0	4	2	0.05[c]
Day 3	0	1	2	0.42[c]

[a]Data denote means (SD).
[b]Analysis of variance.
[c]Chi-square test. For comparison, mean (SD) duration of diarrhea in untreated children ($n = 5$) was 2.6 (1.3) days.

Number of IgA sASC/10^6 Cells to Rotavirus and Mean Levels (95% CI) of Serum IgA Antibodies (EIU) to Rotavirus During the Acute and Convalescent Stage of Rotavirus Gastroenteritis[a]

	LGG	Lactophilus	Yalacta	Analysis of Variance *p*
sASC[b] response				
Acute	0.1 (0–2.3)	0.2 (0–3.1)	0.5 (0–7.2)	0.70
Convalescent	3.3 (0.9–12.7)	0.1 (0–3.2)	0.1 (0–1.5)	0.01
Serum IgA antibodies	(n = 13)	(n = 11)	(n = 12)	
Acute	0.3 (0.1–1.7)	0.3 (0–1.6)	0.1 (0–0.3)	0.41
Convalescent	27.5 (18.9–40)	29.6 (21.1–41.4)	6.1 (1.4–26.1)	0.01

[a]For comparison, sASC response (>1 sACS/10^6 cells) was detected in one of five untreated controls at the acute stage and in none of five at convalescence.

[b]Specific antibody-secreting cells.

Remarks The duration of diarrhea was significantly different between the study groups; those receiving LGG had a shorter duration of diarrhea than those receiving other preparations did. The mean (SD) duration of diarrhea was 1.8 (0.8) days in children who received LGG, 2.8 (1.2) days in those receiving Lactophilus, and 2.6 (1.4) days in those receiving Yalacta ($F = 3.3$, $p = 0.04$). The mean (SD) duration of diarrhea in an untreated comparison group ($n = 5$) was 2.6 (1.3) days. The effect of LGG on the duration of diarrhea was not manifested on the first day; however, after 2 days of treatment, only 19% of patients receiving LGG had diarrheal stools, whereas 64% of patients receiving Lactophilus and 58% of patients receiving Yalacta continued to have watery diarrhea.

The number of IgA sASC to rotavirus at convalescence was higher in patients receiving LGG than in those receiving Lactophilus or Yalacta. Most patients (10 of 11) receiving LGG showed a detectable rotavirus IgA sASC response at the convalescent stage, whereas 3 of 7 patients receiving Lactophilus and 2 of 7 patients receiving Yalacta had a detectable sASC response. All patients had a detectable response in serum rotavirus IgA ELISA antibodies. The mean serum rotavirus IgA antibody levels at convalescent stage were higher in the LGG and Lactophilus groups than in the Yalacta group. All these results confirm that LGG promotes clinical recovery from acute gastroenteritis and potentiates gut immune response to rotavirus.

Experimental Protocol (12)

1. Twenty preterm infants with a gestation age of 33 weeks or less who were resident on a neonatal unit of the Princess Anne Hospital, Southampton, Eng-

land, were studied from the initiation of milk feeds until discharge (September 1, 1991, to January 31, 1992). The infants were randomized to receive either milk feeds or milk feeds supplemented with *Lactobacillus* GG 10^8 CFU twice a day for 2 weeks.

2. Fecal samples were collected each day. Enterobacteriaceae were isolated and enumerated on MacConkey agar. Bacterial isolates were identified by standard biochemical tests. *Lactobacillus* GG was defined as Gram-positive bacilli, catalase negative, forming large white colonies on Rogosa's agar in air after 48 h at 37°C.

Quantitative Changes in Bacterial Population in Fecal Samples from Infants Given *Lactobacillus* GG Supplement for 2 Weeks

	Bacterial Count (Log_{10} CFU/g dry weight)					
		Wk 1	Wk 2			
	Wk 0	Feed Supplement		Wk 3	Wk 4	Wk 5
Lactobacillus GG	4.0[a]	10.4[a]	9.8[a]	6.4[a]	8.7[a]	5.8[a]
Enterobacteriaceae	10.4–11.1	9.2–11.2	9.3–11.3	10.0–11.1	9.4–11.4	9.9–11.1
Enterococci	10.1	8.2–11.9	6.0–11.1	7.8–11.4	8.5–10.9	8.1–10.8
Anaerobes	8.7–10.1	7.0–11.2	8.7–11.4	5.8–11.0	9.1–11.0	10.3–11.0
Staphylococci	6.7–9.8	9.6–10.4	6.7–10.4	7.1–9.0	8.9–10.1	5.5–9.5

[a]Median.

Remarks Orally administered *Lactobacillus* GG was well tolerated and did colonize the intestine of premature infants. However, colonization with *Lactobacillus* GG did not reduce the fecal reservoir of potential pathogens (the bacteria counts of milk-fed group was comparable to the supplement-fed group and data not included). There was no evidence that colonization of *Lactobacillus* GG had any positive clinical benefit: general well-being, abdominal distension, vomiting or regurgitation, incidence of perineal rash, frequency and consistency of stools, number of suppositories used, fluid intake, evidence of sepsis, antibiotic treatment, and any other concomitant medication, oxygen and ventilatory requirements, and duration of hospital stay.

Effect of Streptococci (1)

Profile of Patients

	SF68-Treated Group	*Lactobacillus*-Treated Group	Statistical Analyses
Number of cases	53	51	—
Sex			
Male	26	25	$\chi^2 = 0.04$
Female	27	26	$p > 0.08$
Mean	16.5 mo	20 mo	

(continued)

Profile of Patients (*cont.*)

	SF68-Treated Group	*Lactobacillus*-Treated Group	Statistical Analyses
Range	1 mo–6 y	1 mo–9 y	—
Age			
Up to 6 mo	14	13	$\chi^2 = 0.185$
6–12 mo	14	14	$p > 0.98$
12–24 mo	16	14	
2 y	9	10	
Clinical remarks at the start of treatments			
Duration of diarrhea (days)	3.2	3.1	$t = 0.06$ $p > 0.90$
Bowel movements (average no./day)	4.8	4.6	$t = 0.18$ $p > 0.80$
Mucus in feces (no. of cases)	26	23	$\chi^2 = 0.04$ $p > 0.80$
Blood in feces (no. of cases)	5	4	$\chi^2 = 0.01$ $p > 0.90$
Presence of fever (no. of cases)	25	24	$\chi^2 = 0.03$ $p > 0.80$
Stool cultures (no. of cases)	29	27	$\chi^2 = 0.01$ $p > 0.90$
Duration of cases (in days)			
Means	5.6	5.7	$t = 0.09$ $p > 0.90$
Range	3–9	3–10	
Antibiotic treatment by parenteral route			
Gentamicin	13	12	$\chi^2 = 0.02$ $p > 0.80$
Ampicillin	9	9	
None	31	30	

Experimental Protocol The two coded preparations were packed in identical gelatin capsules, containing, respectively, SF68 preparation (not less than 3.75×10^7 *S. faecium* in lyophilized form) as BIOFLORIN obtained from Giuliani S.A., Lugano, Switzerland, and *Lactobacillus* preparation (*L. acidophilus* 5×10^8, *L. bulgaricus* 5×10^8, *S. lactis* 4×10^9 in lyophilized form). Its contents were taken as a suspension in water. In this double-blind trial, the patients randomly received one of the two preparations, in the following doses: up to 1 year of age (1 capsule b.i.d), 1–3 years (1 capsule t.i.d), 3–7 years (2 capsule b.i.d), and over 7 years (2 capsule t.i.d). Apart from diet and rehydration, no other treatment was given unless respiratory infection was present when systemic antibiotics (gentamicin or ampicillin) were used.

The main clinical parameter used to assess the response to treatment was the presence of diarrhea, which has been defined as the passage of three or more unformed stools per day with or without associated blood and mucus. As criteria of recovery were taken into account the related clinical signs or symptoms such as fever, abdominal pain, metheorism, and the like.

Frequencies of Disappearance and Estimated Persistence Rates

	SF68-Treated Group[a]			*Lactobacillus*-Treated Group		
Periods (days)	N	R	EPP	N	R	EPP
1	53	22	58.5	51	6	88.2
2	31	11	37.7	45	12	64.7
3	20	13	18.2	33	14	37.3
4	7	3	7.5	19	10	17.7
>4	4			9		

[a]N, number of cases at the beginning of each period; R, number of responses (disappearance of the diarrhea) in each period; EPP, estimated percent of the persistence at the end of each period.

Remarks In the patients treated with SF68 preparation, the diarrhea disappeared more rapidly than in the patients treated with the *Lactobacillus* preparation. After two days of treatment, 62.3% of the SF68-treated group had recovered completely compared with only 35.3% in the *Lactobacillus*-treated group. This improvement was independent of age. The SF68 strain seems to have had a specific biological effect that is probably due to the activity on some of the enteropathogens. Thus the use of SF68 strain in the treatment of diarrheal disorders in pediatrics should be considered as an appreciable advance, avoiding significant side effects of other treatments.

Effect of Bifidobacterium *(8)*

Experimental Protocol

1. Thirteen inpatients treated at hospitals connected with the Department of Pediatrics of the School of Medicine, Keio University, Tokyo, Japan, and 2 other patients, were studied, between December 1982 and October 1985.
2. Immediately after collection, the fecal material was put into a test-tube, maintained under anaerobic conditions with CO_2 saturation, and kept at a low temperature (0–5°C). Examination of the fecal specimens was done within 3 h. Modified VL-G medium was used to measure the total number of organisms. Modified VL-G medium, to which 80 μg/ml vancomycin and 1 μg/ml kanamycin were added; MPN medium and CW medium were used to detect *Bacteroidaceae, Bifidobacterium,* and *Clostridium perfringens,* respectively. MPN medium, to which 3000 μg/ml streptomycin and 1000 μg/ml neomycin were added, was used to recover the administered organism, *B. breve.* All these media were prepared by the anaerobic roll tube method. The numbers of *Enterobacteriacea, Enterococcus, Lactobacillus,* and *Staphylococcus* were determined using either DHL medium, KMN medium, or Staphylo No. 110 medium. For the isolation of fungi, candida GS medium was used. In the detection of pathogens from loose stools, *Shigella, Salmonella,* and *Yersinia* were examined using SS medium, *Vibrio cholera* and *V. parahaemolyticus*

with TCBS medium, and *Campylobacter* with Skirrow's medium. At least five strains of *C. difficile* and enteropathogenic *E. coli* were examined from each specimen using CCFA medium and DHL medium, respectively, and the presence or absence of toxin production was examined for each bacteria. Cytotoxin and enterotoxin produced by *C. difficile* were examined using fecal material at 1:10 dilution and culture supernatant as follows: cytotoxin was determined by means of cytopathogenic effects (CPE) on HeLa cells and CPE neutralization by anti-*Clostridium sorderii* antibody; enterotoxin was determined by means of reversed-passive latex agglutination using a kit for enterotoxin detection. Heat-labile toxin (LT) and heat-stable toxin (ST) of enteropathogenic *E. coli* were detected using a commercial kit for enterotoxin detection and by intragastric administration into young mice, respectively.

3. Patients received orally a preparation (BBG-01) containing *B. breve* strain Yakult at 10^9/g, which was isolated from breast-fed healthy infants, and another preparation (BLG-B) containing both *L. casei* strain Shirota at 10^{10}/g and *B. breve* strain Yakult at 10^9/g. In addition to these preparations, some of the patients were given bifidus yogurt that contained *B. breve* strain Yakult, *B. bifidum* strain Yakult, and *L. acidophilus* at 10^{10}/100, 10^{10}/100, and 10^9/100 ml, respectively. These preparations were administered at a mean daily dose of 3 g (divided by 3). A daily amount of 60–600 ml of bifidus yogurt was combined in some of the patients based on their respective clinical states.

Profile of Patients

							Duration of Diarrhea (days)	
Patients No.	Age	Sex	Underlying Disease	Symptoms	Antibiotics	Bacterial Therapy[a]	Before Treatment	After Treatment
1	2 y	F	Periodic granulo-cytopenia	Furuncle	CCL, CET, GM, PIPC	BLG-B	7	7
2	2 y 10 mo	M		Sepsis	KM, GM, ABPC, CBPC, CEZ, CMZ, CTX	BLG-B MIL-MIL	30	14
3	1 y	M	Kawasaki disease	Bronchitis	CEX	MIL-MIL	35	7
4	1 y	M		Sepsis?	LMOX, CET	BBG-01	5	7
5	1 y 8 mo	F	Nephrosis	Salmonellosis	ABPC, ST, KM, FOM	BLG-B	35	8
6	1 mo	M		Bronchitis	ABPC, AMPC	BBG-01	25	7
7	15 y	M	Chronic nephritis	Peritonitis	TOB, CET	BBG-01	11	6
8	15 mo	M	Hirshsprung	Sepsis?	TOB, CBPC, CTX, FOM, Gm, CAZ	BBG-01	10	3

(continued)

Profile of Patients (*cont.*)

Patients No.	Age	Sex	Underlying Disease	Symptoms	Antibiotics	Bacterial Therapy[a]	Duration of Diarrhea (days) Before Treatment	After Treatment
9	1 mo	F	Milk allergy	Sepsis?	ABPC, CEZ, KM	BBG-01	25	7
10	3 y	M	Hemophilia B	Sepsis?	CTX, CLDM, TOB	BBG-01	9	4
11	6 y	M	Hemophilia B	URTI	CTX, CXM	BLG-B	70	10
12	1 mo	M		Sepsis?	ABPC, LMOX	BBG-01	7	4
13	3 mo	M	Ventricular septal defect	Sepsis	CTX, MCIPC, ABPC, LMOX, PIPC, CP	BBG-01	30	10
14	4 y 6 mo	F	Reye syndrome	Broncho-pneumonia	ABPC, PIPC, CET, GM	BBG-01	40	7
15	3 mo	M		Sepsis?	CMZ	BBG-01	40	4

[a]BLG-B, combined preparation of *B. breve* strain Yakult (10^9/g) and *L. casei* strain Shirota (10^9/g), obtained from Yakult Honsha, Tokyo. BBG-01, preparation of *B. breve* (10^9/g), isolated from breast-fed healthy infants. MIL-MIL, bifidobacterium yogurt, containing 10^{10} of viable *B. breve* strain Yakult, *B. bifidum* strain Yakult, and 10^9 of *L. acidophilus* per 100 ml of bottle, produced by Yakult Honsha, Tokyo, under the trade name of MIL-MIL.

Remarks The duration of diarrhea prior to the administration of viable bacterial preparations was an average of 25.3 days (ranging from 5–70 days). Diarrheal passages returned to normal after a mean of 7.0 days (3–14 days) after the administration of bacterial preparation.

Profile of Fecal Bacteria (expressed as log number/gram wet feces)

Patients No.	Bacterio-therapy[a]	Total Bacteria	Bactero-idaceae	*Bifidobacterium* (dosed *B. breve*)	Entero-bacteriaceae	*Entero-coccus*	*Candida*	*Clostridium difficile*[a]	Fecal Toxin CT	ET
1	B	6.46	<5.30	<2.30	<3.30	<3.30	6.46	NT		
	A	10.61	9.43	10.61 (7.85)	9.33	8.99	5.79	NT		
2	B	5.69	<4.30	<2.30	<2.30	<2.30	5.69	NT		
	A	9.75	8.23	9.49 (9.36)	8.43	8.09	3.41	NT		
5	B	9.72	<4.30	<2.30	4.66	9.72	7.16	<3.30		
	A	10.23	9.81	9.81 (9.15)	9.94	8.08	5.76	<3.30		
6	B	7.40	7.25	<1.81	7.39	3.77	<2.81	<2.81		
	A	10.26	8.58	10.15 (9.76)	8.48	7.60	<2.32	<2.32		
7	B	9.55	7.85	8.86	7.75	5.67	<2.30	<2.30		
	A	10.51	7.00	10.04 (9.00)	6.56	7.08	<2.30	<2.30		
8	B	9.04	9.00	<2.30	7.34	7.43	7.79	<2.30		
	A	10.34	7.77	9.88 (9.88)	9.23	8.48	2.30	NT		

(continued)

Profile of Fecal Bacteria (expressed as log number/gram wet feces) (*cont.*)

									Fecal Toxin	
Patients No.	Bacteriotherapy[a]	Total Bacteria	Bacteroidaceae	*Bifidobacterium* (dosed *B. breve*)	Enterobacteriaceae	*Enterococcus*	*Candida*	*Clostridium difficilea*	CT	ET
9	B	9.29	8.92	6.32	9.29	<2.67	<2.67	<3.67		
	A	10.55	9.38	10.53 (9.77)	9.53	7.68	<2.34	<3.34		
10	B	9.62	8.72	5.91	9.62	6.48	<2.30	<3.30		
	A	10.08	9.87	9.89 (9.60)	8.97	8.68	<2.30	<3.30		
11	B	9.58	8.06	6.31	9.56	8.33	5.39	<3.30		
	A	10.68	9.11	10.26 (9.93)	9.06	10.00	5.19	<2.30		
12	B	5.26	<1.30	<1.30	5.26	<2.30	<2.30	<2.30		
	A	10.32	5.82	10.32 (8.73)	8.83	5.82	<2.30	<2.30		
13	B	8.15	<3.96	<2.96	<2.96	3.45	8.15	<2.96	–	–
	A	10.19	9.36	9.26 (9.26)	9.71	9.89	<2.96	6.57	+	+
14	B	10.52	<8.29	<5.67	8.85	9.63	4.35	2.44	–	–
	A	10.58	8.51	10.35 (7.20)	9.19	9.48	5.60	7.68	–	–
15	B	10.38	<6.30	<4.30	10.04	9.87	<2.30	<2.30	–	+
	A	10.18	9.49	9.66 (9.30)	9.61	9.04	<2.30	<2.30	–	+

[a]B, before bacteriotherapy; A, after bacteriotherapy; NT, not tested.

Remarks The total number of bacteria before the treatment had clearly decreased, to 8.51 (1.77 /g (mean log value ± SD per 1 g wet feces weight), but this number returned to the normal level of 10.21 ± 0.26 /g during the treatment, when diarrhea was cured. The most dominant bacteria before treatment included facultative anaerobes (*E. coli* group in 6 cases, *Enterococcus* in 2 cases, and *Candida* in 3 cases). *Bifidobacterium* were not detected in 9 of the 13 patients; the number of cells was decreased to 6.97 ± 1.31 /g even in the four patients with *Bifidobacterium,* indicating eradication or marked decreases of *Bifidobacterium.* After diarrhea was cured by treatment with viable bacterial preparations, *Bifidobacterium* predominated over other bacteria in the intestinal tract in 11 of the 13 patients. Bifidobacterium together with other bacteria (*E. coli* group and *Enterococcus* in one patient each) predominated in the remaining 2 patients, with the number of *Bifidobacterium* returning to the normal level of 9.82 ± 0.96 /g. The administered *B. breve* was recovered from all of the 13 patients, at a level of 9.68 ± 0.96/g.

From these results, decreases in anaerobes, especially decreases or eradication of *Bifidobacterium,* were markedly observed during the diarrhea period (before treatment), while increases in facultative anaerobes, including the *E. coli* group, *Enterococcus,* and *Candida* were characteristic of this period. After diarrhea was cured by treatment with vital bacterial preparations, normal flora, consisting of *Bifidobacterium* for the most part, was formed. Since *B. breve* was a component common to the three kinds of viable bacterial preparations and it predominated among the administered bacteria in all the patients, it was postulated to play the most important role in the cure of diarrhea.

Prevention of Traveler's Diarrhea (15)

Experimental Protocol

1. The study subjects were 820 travelers (age 10 to 80 years, mean age 43.8 ± 13.9) traveled on holiday to two resort towns on the southern Mediterranean coast of Turkey. 528 of them traveling to Marmaris (350 for 1 week and 178 for 2 weeks) and 228 to Alanya (147 for 1 week and 81 for 2 weeks). They were randomized into two groups receiving either *L. rhamnosus* GG powder (402 persons), provided by Valio Finnish Co-operative Dairies Association, Helsinki, or a placebo containing ethyl cellulose powder (418 persons) packed in identical sachets. The subjects were instructed to take the contents of one sachet mixed with cold water twice daily for 2 days before departure and continuing them during the trip. The daily dose of *Lactobacillus* GG was about 2×10^9 bacteria. A Finnish physician was available during the trip at both destinations to register the cases of diarrhea, to observe side effects, and to give medical treatment when necessary.
2. No living, eating, or drinking restrictions were placed on the participants during the study. The weather in both resorts and the countryside was similar and typical for that time of year; daytime temperatures reached 35°C.
3. Diarrhea was defined as three or more unformed stools lasting more than 24 h or one to two unformed watery stools in less than 24 h.

Relation Between Traveler's Ages and Incidence of Diarrhea

	Marmaris		Alanya	
Age Group (y)	Placebo (%)	*Lactobacillus* GG (%)	Placebo (%)	*Lactobacillus* GG (%)
10–29	52.6	47.4	60.0	7.1
30–39	41.8	40.0	57.9	22.7
40–49	32.8	39.5	35.1	39.5
50–59	45.8	28.3	26.1	33.3
60–80	28.3	42.4	23.1	9.1

Remarks Increasing age was negatively correlated with the incidence of diarrhea in the placebo group in Marmaris. The differences in the effectiveness of *Lactobacillus* GG in the two different locations could be due to the presence of different pathogens in the two places. Another possible contributing factor for the apparent ineffectiveness of *Lactobacillus* GG in Marmaris might have been the uneven age distribution of the Marmaris placebo group in which older people were very well represented. *Lactobacillus* GG was well tolerated in all subjects and no side effects could be demonstrated. The study suggests that *Lactobacillus* GG can be safely administered to healthy people in doses of 2×10^9 bacteria for up to 2 weeks. Its use can diminish the risk of traveler's diarrhea during trips abroad, although the results are not uniform in all locations.

Protection Against Antibiotic-Associated Diarrhea

Effect of Lactobacillus *(6)*

Experimental Protocol This study was performed during the 15-month period from February 1977 to April 1978 and involved 79 patients in all. After identification of a patient who had started ampicillin therapy, the candidate was evaluated for entry into the study if he or she had diarrhea with three or more bowel movements, regardless of consistency. Patients with diarrhea were followed until their symptoms resolved and were classified according to the most likely etiology: category 1—diarrhea caused by diet or medications other than ampicillin; category 2a—diarrhea caused by ampicillin; category 2b—diarrhea caused by ampicillin or other medications, or both; category 2c—diarrhea caused by ampicillin or underlying disease, a combination of the two or other medications. In a double-blind procedure, the patients were assigned randomly to receive one packet of Lactinex (*L. acidophillus* and *L. bulgaricus*) or placebo four times daily for the first 5 days of ampicillin therapy.

Incidence and Probable Etiology of Diarrhea

	Placebo		Lactinex[a]	
No. of patients	43		36	
Male	22		13	
Female	21		23	
No. of patients with diarrhea (% total)	9 (21%)		3 (8.3%)[b]	
Category 1	3		3	
Category 2a[c]	3	(14%)	0	(0%)[d]
Category 2b[c]	1		0	
Category 2c[c]	2		0	

[a]Contained *L. acidophilus* and *L. bulgaricus,* obtained from Hynson, Westcott, and Dunning, Baltimore, MD.
[b]Difference in proportions not significant ($p = 0.21$, Fisher's Exact Test).
[c]Diarrhea associated with ampicillin.
[d]Significant difference in proportions ($p = 0.03$, Fisher's Exact Test).

Remarks There were no significant differences ($p > 0.05$, one-way analysis of variance) in patient age, dose of ampicillin, or number of doses of the study drug in the two groups. However, there were more women than men in the study population. Diarrhea occurred in 9 of 43 patients (21%) receiving placebo as compared with 3 of 36 patients (8.3%) receiving Lactinex. This difference was not statistically significant. When diarrhea was classified according to its most likely etiology, causes other than ampicillin were implicated in 50% of the patients. These causes included the patients' diet and medications other than ampicillin. When these conditions were excluded, the incidence of diarrhea caused by ampicillin was 14% in the placebo group and 0% in the Lactinex group. This difference was statistically significant. Thus the prophylactic ingestion of a commercial preparation of lactobacilli (Lactinex) may be effective in preventing ampicillin-associated diarrhea in adults.

Effect of Streptococcus faecium *(3)*

Experimental Protocol The study was carried out in 190 inpatients admitted to "S. Biagio" and MOA Locatelli Hospital, Bergamo, Italy because of pulmonary chronic tuberculosis. Antitubercular therapy with different combinations of antitubercular drugs (streptomycin, isoniazid, ethambutol, rifampin, *p*-aminosalicyclic acid, kanamycin, and cycloserine) were started in these patients. At the same time, patients with normal bowel habit were randomly assigned according to a "between subjects" design, to one of the following groups: control group (C) without other treatments; reference (R) treated also with vitamins (A 2000 UI, B_1 3 mg, B_2 3 mg, B_6 3mg, D_2 200 UI, E 8 mg, B_{12} 3.2 mg, C 500 mg, PP 10 mg); groups T1, T2, and T3, treated respectively with 1, 2, and 3 capsules/day containing at least 75 million *S. faecium* bacteria SF 68 strain (SF 68) in lyophilized form. This study lasted 2 months. In order to assess the properties of SF 68 to prevent possible antibiotic reactions, the parameters evaluated were as follows: appearance of episodes of diarrhea, defined as more than two bowel movements/day of watery stools that went on for at least 2 days, and signs of vitamin deficiencies such as stomatitis or glossitis.

Description of the Compared Groups

Group	C	R	T1	T2	T3	Total
No. of cases	38	37	39	38	38	190
Sex						
Male	19	18	19	20	19	95
Female	19	19	20	18	19	95
Age						
Mean (y)	61	61	56	62	58	59
Range	22–87	29–89	19–71	19–81	23–76	19–89
No. of patients taking the following antibiotics:						
Streptomycin	15	11	16	11	16	69
Isoniazid	35	33	38	35	35	176
Ethambutol	36	37	39	37	32	183
Rifampin	32	18	26	32	30	138
p-aminosalicyclic acid	1	3	2	3	4	13
Kanamycin	1	2	4	1	2	10
Cycloserine	2	3	3	2	2	12

Frequencies of Cases Showing Episodes of Diarrhea or Signs of Mucositis in the Compared Groups[a]

Groups	C	R	T1	T2	T3	Statistical Analysis		
Diarrhea						*Overall*	$\chi^2 = 12.0$	$p < 0.02$
Frequency	7	4	2	0	1	C + R vs. T	$\chi^2 = 9.67$	$p < 0.01$
Percent	18	11	5	0	3	C vs. R	$\chi^2 = 1.59$	$p > 0.20$
						T1 vs. T2 vs. T3	$\chi^2 = 0.74$	$p > 0.50$

(continued)

Frequencies of Cases Showing Episodes of Diarrhea or Signs of Mucositis in the Compared Groups[a] (*cont.*)

Groups	C	R	T1	T2	T3	Statistical Analysis		
Mucositis						*Overall*	$\chi^2 = 11.11$	$p < 0.05$
Frequency	5	3	0	1	0	C + R vs. T	$\chi^2 = 9.66$	$p < 0.01$
Percent	13	8	0	3	0	C vs. R	$\chi^2 = 1.05$	$p > 0.30$
						T1 vs. T2 vs. T3	$\chi^2 = 0.40$	$p > 0.80$

[a]C, control group (antibiotics only); R, reference group (antibiotics plus vitamins); T1, T2, T3, treated groups (antibiotics plus, respectively 1, 2, and 3 capsules/day of the SF 68 preparation).

Remarks The groups were comparable in number, sex, age, and pattern of employed antibiotics. The rates of patients who had diarrhea as well as those who showed signs of mucositis as a possible consequence of vitamin deficiencies in the control group and reference group were not very high. However, the test treatment (T1, T2, and T3 groups) exhibited a highly significant activity in preventing diarrhea and in avoiding mucositis; only 3 cases out of 115 had episodes of diarrhea and 1 had an oral mucositis. Thus the SF 68 administration is useful in preventing diarrhea and development of mucositis, and this could be explained by the maintenance of a normal bowel microecology due to the already well-known properties of the SF 68 strain to prevent possible alterations induced by antibiotics or other factors.

Treatment of Relapsing C. difficile *Colitis*

Experimental Protocol (5) The patients had had two to five relapses over a 2–10 month period. Each relapse was associated with diarrhea, abdominal pain, and weight loss. Tests for *C. difficile* cytotoxin in the stool were positive during relapses. Previous therapies had provided temporary relief, but clinical and microbiological relapse occurred within 10 days. *L. rhamnosus* strain GG isolated from human, was given as a concentrate in 5 ml of skim milk, at a daily dose of 10^{10} viable bacteria, for 7–10 days. In patients A, B, and C, *Lactobacillus* GG treatment was started just after completing a 10-day course of antibiotics (metronidazole or bacitracin), while in the other two, GG was given during the active relapse phase, at least 2 weeks after the previous antibiotic treatment.

Details of Patients with Relapsing *C. difficile* Colitis

Patient (age/sex)	Inciting Antibiotic	No. of Relapses	Duration of Illness (mo)	Previous Therapies[a]	Toxin Titers		Follow-up with No Relapses
					Before	After	
A (35/M)	Cephalexin	2	2	Met	1/1250	0	16 mo
B (93/M)	Erythromycin	3	4	Met, Vanco	1/1250	0	1 y
C (24/F)	Clindamycin	3	3	Met, Vanco, Baci	1/1250	0	4 y
D (73/F)	Ampicillin	4	6	Met, Vanco, Baci	1/1250	1/10 delayed	4 mo
E 50/M)	Penicillin	5	10	Met	Positive (no titre)	0	4 y

[a]Met, metronidazole; Vano, vancomycin; Baci, bacitracin.

Remarks With *Lactobacillus* GG treatment, patients A, B, C, and E had an immediate satisfactory result, with termination of diarrhea, no further relapses, and negative, or very low toxin titers in the stools. Patient C received *Lactobacillus* GG for 7 days and had an initial improvement but relapsed 3 days later with diarrhea and positive toxin titers. She was treated with metronidazole for 10 days, followed by *Lactobacillus* GG, after which she had no further relapses. These results suggest that *Lactobacillus* GG is efficacious in terminating relapsing colitis due to *C. difficile.*

Experimental Protocol (2) Four children had had three to five relapses each of *C. difficile* colitis. They had been treated with several courses of vancomycin (40 mg/kg/day) and metronidazole (20–30 mg/kg/day). In three patients, the bloody diarrhea and cramping improved during treatment with these medications but recurred within several weeks of the end of therapy. Patient 2 continued to have bloody diarrhea despite these medications. All patients tested positive on *C. difficile* cytotoxin assay during and after each course of treatment. Three patients were healthy prior to the onset of the diarrheal illness; patient 3 had previously documented gastroesophageal reflux, formula intolerance requiring a hypoallergenic formula, delayed gastric emptying following rotavirus illness, and hypogammaglobulinemia. *C. difficile* cytotoxin assays (toxin A and B) were positive initially and during relapses in all patients. All patients had bloody diarrhea and abdominal pain prior to onset of treatment with *L. rhamnosus* strain GG (provided by Valio Dairies, Helsinki, Finland), which was prepared as a freeze-dried powder at a concentration of 5×10^9 viable bacteria per gram. It was given in a dose of 125 mg orally b.i.d. for 2 weeks. No other antimicrobial drugs were administered during this period; however, antibiotics were subsequently used in all patients during the follow-up period. *C. difficile* cytotoxin assays were performed prior to and monthly after treatment.

***Lactobacillus* GG Treatment for Relapsing *C. difficile* Colitis**

Patient No.	Age	No. of Relapses	Duration of Illness (mo)	Previous Treatment	Follow-up with No Relapse (mo)
1	4 y	3	3	Vancomycin ×3; metronidazole ×1	18
2	11 mo	4	5	Vancomycin ×2; metronidazole ×2	9
3	5 mo	5	5	Vancomycin ×3; metronidazole ×3	8
4	5.5 y	3	4	Vancomycin ×3; metronidazole ×1	10

Reproduced with permission of Lippincott Williams & Wilkins.

Remarks All patients responded clinically within 5–7 days, with a marked decrease in stool frequency and cramping. All patients were asymptomatic at the end of the 2-week treatment period and were cytotoxin-negative on assay; however, patients 1 and 3 became toxin-positive, with recurrence of diarrhea, within 1–2 months and were retreated. Both of these patients became asymptomatic and remained toxin-negative. During the follow-up period (8–18 months; mean 11), all patients remained asymptomatic with negative *C. difficile* cytotoxin assays despite retreatment with antibiotics. No side effects were noted during the course of treatment with *Lactobacillus* GG.

Preservation of Intestinal Integrity during Radiotherapy (18)

Experiment Protocol

1. This study included 21 patients with the diagnosis of cervix or uterus carcinoma. All subjects were scheduled for internal and external pelvic radiotherapy starting with intracavitary cesium administration. After an interval of 1–2 weeks a Wertheim hysterectomy was performed. After 4–6 weeks, the pelvis was irradiated using a 15 × 15 cm anterior and posterior portals on megavoltage equipment. The prescribed dose was 4400 cGy midline dose in 22 fractions daily in a split course; there was a 2-week break in the middle of treatment. The calculated sum of internal and external radiation was 8000 cGy for the tumor and 5000 cGy for the pelvic area.
2. The subjects were 40–75 years old and of relatively similar body mass index. Diabetics and subjects with gastrointestinal disorders were excluded. Patients fulfilling these criteria were randomized into two groups. The control group received dietary counseling only. The test group received both dietary counseling and a daily dose of at least 2×10^9 live *L. acidophilus* (NCDO 1748, National Collection of Dairy Organisms, Reading, UK) bacteria in the form of a yogurt-type product. To support the growth of the *L. acidophilus* in the large intestine 6.5% lactulose was added. The radiation stability of the acidophilus strain was tested using a radiation dose 10 times greater than that given to the patients.
3. The test group received 150 ml of the product daily for 5 days prior to radiotherapy, daily throughout the radiotherapy period including the interval, then for 10 days after finishing the therapy regimen. The control group received only dietary counseling. Dietary advice was planned to emphasize sufficient energy and protein intake. Increased fluid intake was recommended if diarrhea occurred. The patients were also recommended to consume small meals to avoid nausea and diarrhea. Antidiarrheal drugs were prescribed as required.
4. All patients were interviewed regularly before treatment (control time 1), during the treatment (control times 2, 3, and 4) and 6 weeks after the treatment (control time 5) by a physician. Weight was recorded at frequent intervals and results were analyzed statistically using Fisher's exact test.

	Control Group ($n = 10$)					Test Group ($n = 11$)				
	No. Patients with Side Effects at Control Time					No. Patients with Side Effects at Control Time				
	1	2	3	4	5	1	2	3	4	5
Diarrhea	0	8	9	8	9	1	3[a]	2[a]	2[a]	3[a]
Loss of appetite	1	3	2	2	1	1	2	3	2	4

(continued)

	Control Group ($n = 10$)					Test Group ($n = 11$)				
	No. Patients with Side Effects at Control Time					No. Patients with Side Effects at Control Time				
	1	2	3	4	5	1	2	3	4	5
Flatulence	3	3	5	5	7	4	7	9	7	6
Mean weight	68.2				67.1	67.3				65.8
(± SD)	12.5				12.3	14.6				14.5
Weight loss[b]					5					5

[a]Significantly different from the control group ($p < 0.01$).
[b]Number of patients losing at least 1 kg during treatment.

Remarks All subjects in the control group suffered from diarrhea during the radiotherapy. Diarrhea was pronounced and lasted until the treatment was completed. Most control patients had several episodes of diarrhea in spite of the dietary advice. The patients consuming a daily dose of yogurt had only occasional cases of diarrhea during the study and diarrhea was of a transient nature. The incidence of diarrhea was significantly smaller in the yogurt group than in the control group ($p < 0.01$). There were no differences in the incidence of vomiting, nausea, abdominal pain, loss of appetite, or weight loss between the groups. However, the yogurt group experienced more flatulence than the controls. The control patients used more antidiarrheal drugs (six permanent users) than the yogurt group (one permanent diarrheal user and one laxative user). The results of this study therefore indicated that the consumption of a yogurt product containing live *L. acidophilus* cultures and lactulose as a supply for the bacteria significantly decreased the incidence of radiotherapy-induced diarrhea in patients receiving pelvic area radiotherapy.

Protection Against Lethal Irradiation (14)

Experimental Protocol and Assay Methods

1. Animals: Male C3H/HeN mice (9–10 weeks old) or female BALB/c mice (15 weeks old) were used. They were housed randomly, five mice per cage, checked daily, and surviving animals were killed by cervical dislocation 31 days after irradiation.
2. Irradiation: 10 nonanaesthetized mice at a time were exposed to whole-body radiation from a source of ^{137}Cs γ rays while they were individually restrained. The exposure rate was monitored by a Victoreen R meter and a factor of 0.956 was used for conversion of roentgens to cGy.
3. *L. casei*: Lyophilized preparation of *L. casei* strain Shirota, which were killed by heating at 100°C for 30 min, was stored at –20°C. *L. casei* strain Shirota was obtained from Yakult Central Institute for Microbiological Research, Tokyo. The bacterial preparation was suspended in pyrogen-free saline at densities of 1–10 mg dry weight/ml and injected in 0.1-ml portions subcutaneously into the left inguinal region of the mice. It was given in a single dose

within 10 min after irradiation or in divided doses (one daily for 4 days). In the latter case, the first injection was given immediately after irradiation. Control animals were given 0.1 ml saline at the same time.

4. Hematological methods: Blood was collected by heart puncture under ether anesthesia. The number of leukocytes, erythrocytes, and platelets in individual animals was scored with an automatic hemocytometer. The spleen and both femurs were excised from the killed mice, and the number of granulocyte-macrophage colony-forming cells (GM-CFC) was assayed. Briefly, samples of spleen and bone marrow were made into single-cell suspensions, pooled for each group of five mice, and cultured in triplicate. The cell density at the start of culture was either 2.5×10^5 (spleen) or 5×10^4 (bone marrow) per dish. McCoy's 5A medium was supplemented with 0.32% agar, 20% horse serum (GIBCO), and 200 units/ml of recombinant mouse granulocyte-macrophage colony stimulating factor (GM-CSF). Plastic petri dishes were used and the final volume of the culture was 1 ml per dish. Incubation was carried out at 37°C in 5% CO_2 /air and colonies (> 50 colonies) developed in 7 days were scored using an inverted microscope. In either experiment, blood was collected from the carotid artery under ether anesthesia and serum was separated after coagulation at room temperature. The serum was heated at 56°C for 30 min to inactivate complement, appropriately diluted, and subjected to the CSF assay. Bone marrow cells were cultured as above except for the use of recombinant GM-CSF. Colony-stimulating activity (CSA) was expressed as colonies/milliliter serum based on activity to give 40–70 colonies/dish, about a half-maximum reaction.

Serum Colony-Stimulating Activity after Irradiation

Effect of *L. casei* Strain Shirota (LC 9018) on Serum CSA in Irradiated C3H/HeN Mice

Hour after Irradiation	Colony-Stimulating Activity (Colonies/ml serum) in Irradiated C3H/HeN Mice Treated with	
	LC 9018[a]	Saline[a]
1	250	15
3	225	200
6	535	175
9	575	560
12	625	15
18	575	20
25	535	125

[a]Immediately after administration of 8.5 Gy, the animals were injected with 1 mg LC 9018 or with saline and killed on the days indicated. Each value represents five animals.

Remarks Colony-stimulating activity had increased to a detectable level in serum 3 h following lethal (8.5 Gy) irradiation, while it was too low to be detected in the serum of normal mice. The serum CSA reached a maximum 9 h after irradiation and then decreased rapidly in the irradiated, saline-treated animals. Colony-stimulating activity was detectable as early as 1 h after irradiation in the serum of LC 9018-treated mice. A maximum level of CSA was reached at 6 h and was maintained for 18 h or longer. Morphological examination showed that the colonies stimulated by the serum of irradiated mice were composed of granulocyte/macrophages or of granulocytes only irrespective of whether the irradiated animals were treated with LC 9018.

Number of GM-CFC in Femur and Spleen

Effect of LC 9018 on Number of GM-CFC in Femur and Spleen[a]

	Number of GM-CFC in					
	Femur			Spleen		
Days after Irradiation	Saline (NR)	LC 9018 (R)	Saline (R)	Saline (NR)	LC 9018 (R)	Saline (R)
3	9.5×10^3	5×10^1	3×10^1	6×10^3	0	0
7	9.5×10^3	6×10^2	0	6×10^3	1×10^1	0
10	9.5×10^3	5×10^2	0	6×10^3	5×10^2	0
14	9.5×10^3	5×10^3	4.5×10^1	6×10^3	1.6×10^4	1×10^1
17	9.5×10^3	6×10^3	3×10^2	6×10^3	7.5×10^4	1×10^3
21	9.5×10^3	1.6×10^4	5×10^3	6×10^3	1×10^5	6×10^4

[a]Each group consisted of 30 C3H/HeN mice, and five mice from each group were killed on the days indicated. Each value represents the result of a triplicate assay of the pooled marrow or spleen cells. Saline (NR), nonirradiated mice receiving saline; LC 9018 (R), mice receiving 7.5 Gy radiation and 1 mg LC9018; saline (R), mice receiving 7.5 Gy radiation and saline. The difference between 7.5 Gy + saline and 7.5 Gy + LC 9018 was significant ($p < 0.01$) for both femur and spleen.

Remarks The spleen contained fewer GM-CFC than the femoral bone marrow in normal mice. After total-body sublethal irradiation, the number of GM-CFC decreased markedly in both femur and spleen and showed no regeneration until day 10. In contrast, LC 9018 in the irradiated mice significantly accelerated the recovery of GM-CFC number. Although regeneration began a few days earlier in the femur than in the spleen, the number of spleen GM-CFC increased more rapidly than femoral GM-CFC and was 20 times higher by day 21 than before irradiation. This spleen GM-CFC population was 10 times the size of the femoral GM-CFC population in normal mice.

Leukocytes, Erythrocytes, and Platelets in Peripheral Blood

Effect of Treatment with LC 9018 on Number of Blood Erythrocytes (RBC), Platelets (PLT), and Leukocytes (WBC) in Irradiated C3H/HeN Mice[a]

		Days after Irradiation					
		3	7	10	14	17	21
RBC ($\times 10^{-5}$/mm^3)	Saline (NR)	87	87	87	87	87	87
	LC 9018 (R)	70	66	54	48	44	54
	Saline (R)	64	54	50	33	28	23

(*continued*)

Effect of Treatment with LC 9018 on Number of Blood Erythrocytes (RBC), Platelets (PLT), and Leukocytes (WBC) in Irradiated C3H/HeN Mice[a] (*cont.*)

		Days after Irradiation					
		3	7	10	14	17	21
PLT ($\times 10^{-4}/mm^3$)	Saline (NR)	96	96	97	100	88	93
	LC 9018 (R)	85	15	6	6	13	60
	Saline (R)	80	8	0	0	5	8
WBC ($\times 10^{-2}/mm^3$)	Saline (NR)	33	20	25	25	20	24
	LC 9018 (R)	6	6	6	10	29	125
	Saline (R)	0	0	0	0	0	15

[a]Each value represents the mean of 5 mice. Saline (NR), nonirradiated mice receiving saline; LC 9018 (R), mice receiving 7.5 Gy radiation and 1 mg LC 9018; saline (R), mice receiving 7.5 Gy radiation and saline. The difference between 7.5 Gy + saline and 7.5 Gy + LC 9018 was significant ($p < 0.01$) on day 21 for erythrocytes, platelets, and leukocytes.

Remarks Circulating leukocytes (WBC) decreased in number in the LC 9018-treated mice as markedly as in the irradiated control mice and remained low for nearly 2 weeks. The number of leukocytes in the treated mice started to increase on day 14, exceeded the normal level a few days thereafter, and become five to six times greater than normal on day 21. The decrease in platelets lagged several days behind that of leukocytes and occurred in the same fashion in the LC 9018-treated mice as in the controls. The number of platelets started to increase on day 17 in the treated mice. The effect of LC 9018 was statistically significant at the 1% level on day 21, although recovery had progressed only to about 60% of normal levels by this time after irradiation. Erythrocytes (RBC) were also affected by the LC 9018 treatment; blood erythrocyte number continued to decrease on day 17 after 7.5 Gy irradiation in the saline-treated mice, whereas it started to increase in the mice treated with LC 9018. The difference between the saline-treated and the LC 9018-treated was significant at the 1% level on day 21. Erythrocytes were the least affected as indicated in the table.

Influence of Dose of Radiation on Radioprotective Effect of LC 9018

	30-Day Survival (%)[a]	
Radiation Dose (Gy)	Saline Treated	LC 9018 Treated[b]
7.5	80	NT[c]
8.0	50	100[d]
8.5	20	100[d]
9.0	0	90[d]
9.5	NT	20
10.0	NT	0

[a]Each group consisted of 10 mice of the C3H/HeN strain.
[b]LC 9018 (1 mg/mouse) was injected in a single dose immediately after irradiation.
[c]NT, not tested.
[d]Statistically significant ($p < 0.01$).

Remarks The dose of radiation affects the efficacy of radioprotection by LC 9018. At 8 Gy in C3H/HeN mice, it was possible to save all mice by treatment with 1 mg LC 9018, whereas no mice survived after 10 Gy despite the treatment with 1 mg LC 9018. It was calculated from the results shown in the table that the dose of radiation that was lethal for 50% of the mice within 30 days ($LD_{50/30}$) was 7.97 Gy (95% confidence limit, 7.64–8.23 Gy) for the control mice and 9.30 Gy (95% confidence limit, 9.11–9.48 Gy) for the mice treated with LC 9018. Thus the dose reduction factor afforded by LC 9018 was about 1.2 at most.

Effect of Timing of LC 9018 Administration on Mortality of Irradiated C3H/HeN Mice (19)[a]

Group	Day of LC 9018 Injection	30-Day Survival (%)
	–7 days	30
	–3 days	60
I	–2 days	96
I	–1.5 days	100
I	–1 day	100
I	–15 h	100
I	–9 h	90
I	–6 h	90
	0 h	74
II	+ 0.1 h	87
II	+ 3 h	93
II	+ 6 h	80
II	+ 9 h	96
	+ 12 h	90
	+ 30 h	30
	+ 3 days	0

[a]LC 9018 (heat-killed *L. casei* strain Shirota obtained from Yakult Central Institute for Microbiological Research, Tokyo, 1 mg in 0.1 ml saline) was injected in a single subcutaneous dose on various days before (–) and after (+) irradiation, as indicated. The 30-day survival rate was 7% for the 100 control animals that received saline immediately after 8.5 Gy irradiation. Each value represents values for 20–100 mice.

Remarks In group I, mice received a single subcutaneous LC 9018 injection 6, 9, 15, 24, 32, or 48 h before 8.5 Gy total-body irradiation, an average of 96% survived for 30 days after irradiation. Whereas 93% of the mice in the control group that had received the saline injection died within 20 days. Compared with group I mice, an average of 89% of the mice in group II, which received a LC 9018 injection 0.1, 3, 6, or 9 h after irradiation survived. The difference between group I and group II was significant ($p < 0.05$), indicating that LC 9018 was somewhat more effective when used before than after irradiation. The survival rate in a group of mice that received LC 9018 for 3 days before irradiation was 60%. Although this rate was significantly larger than that in the control saline-treated group ($p < 0.01$), the life-saving effect of LC 9018 in this group was less marked than its effect in group I ($p < 0.01$). The effect of LC 9018 was further reduced when it was given 7 days before or 30 h after irradiation. These findings indicate that to achieve radioprotection with LC 9018, the timing of administration before or after irradiation is limited. Nevertheless, the increased survival rate in 7-day preirradiation or 30-h postirradiation treated mice was still significantly better ($p < 0.01$ or 0.05, respectively) than the survival rate (7%) in the saline-treated mice.

Prevention of **Helicobacter pylori** *Infection (10)*

	Weeks after Treatment			
	1	3	4	6
Germ-free mice[a]				
H. pylori (log_{10} CFU/g tissue)[b]	5.31 ± 0.07	4.91 ± 0.07	—	5.19 ± 0.07
Antibody title (492 nm)[c]	0	0.63 ± 0.25	—	0.42 ± 0.16
Other bacteria (log_{10} CFU/g tissue)	0	0	—	0
L. salivarius mono-associated mice[d]				
H. pylori (log_{10} CFU/g tissue)[b]	0	0	—	0
Antibody title (492 nm)[c]	0	0	—	0
L. salivarius (log_{10} CFU/g tissue)[d]	8.26 ± 0.07	9.02 ± 0.03	—	8.59 ± 0.07
Murine stomach origin lactobacilli-associated mice				
H. pylori (log_{10} CFU/g tissue)[b]	0	0	—	0
Antibody title (492 nm)[c]	0	0	—	0
Lactobacilli (log_{10} CFU/g tissue)[d]	9.23 ± 0.21	9.30 ± 0.14	—	8.80 ± 0.07
E. faecalis mono-associated mice[e]				
H. pylori (log_{10} CFU/g tissue)[b]	4.97 ± 0.03	4.86 ± 0.07	—	5.08
Antibody title (492 nm)[c]	0	0.14	—	0.29 ± 0.12
E. faecalis (log_{10} CFU/g tissue)[e]	7.71 ± 0.14	5.77 ± 0.78	—	7.14 ± 0.71
S. aureus mono-associated mice				
H. pylori (log_{10} CFU/g tissue)[b]	5.08 ± 0.03	4.91 ± 0.07	—	5.08 ± 0.03
Antibody title (492 nm)[c]	0	0.14	—	0.28 ± 0.12
S. aureus (log_{10} CFU/g tissue)[f]	6.73 ± 0.14	4.40 ± 0.07	—	6.91 ± 0.78
H. pylori infected gnotobiotic mice[b]				
H. pylori (log_{10} CFU/g tissue)[b]	—	—	5.13 ± 0.12	—
Antibody title (492 nm)[c]	—	—	4.19 ± 1.46	—
H. pylori infected gnotobiotic mice + *L. salivarius*[g]				
H. pylori (log_{10} CFU/g tissue)[b]	—	—	2.60 ± 0.15[h]	—
Antibody title (492 nm)[c]	—	—	2.14 ± 0.02[i]	—

[a]Male germ-free (GF) BALB/c mice were obtained from Nippon CLEA Inc. (Tokyo, Japan).
[b]*H. pylori* 130 (*cag A*$^+$, vacuolating toxin$^+$) was isolated from gastric biopsy materials of patients treated at Tokai University Hospital. For infection with *H. pylori*, 5-week old mice were orally inoculated on three consecutive days with 1×10^9 CFU of *H. pylori* that were freshly cultured in Brucella broth (Difco Laboratories) containing 5% fetal calf serum, in 5% O_2, 10% CO_2, and 85% N_2, at 37°C for 72 h, and re-suspended in 0.5 ml of phosphate buffered saline. The CFU of the bacterium was counted on Skirrow plates.

[c]Serum titer of IgG antibody raised to *H. pylori* was measured by ELISA. The antigen was a bacterial sonicate obtained by ultracentrifugation of *H. pylori.*
[d]*L. salivarius* WB 1004 was provided by Wakamoto Pharmaceutical Co., Tokyo, Japan. The number of CFU of lactobacilli was counted on BL agar (Nissui, Tokyo, Japan). For oral inoculation into mice, lactobacilli were grown in MRS broth (Difco) in 10% H_2, 10% CO_2, and 85% N_2 for 1 day.
[e]*Entercoccus faecalis* 19433 was obtained from ATCC and grown on BHI agar plates.
[f]*Staphylococcus aureus* 25923 was obtained from ATCC and grown on BHI agar (Difco) plates.
[g]GF mice infected with *H. pylori* at 5 weeks old were inoculated orally once a day for three consecutive days at 9 weeks old and thereafter once a week at 10, 11, and 12 weeks of age with 1×10^8 *L. salivarius,* and were sacrificed at 13 weeks of age for examination.
[h]$p < 0.01$.
[i]$p < 0.05$.

Remarks *H. pylori* could not colonize the stomach of *L. salivarius* infected gnotobiotic BALB/c mice but colonized in large numbers and subsequently caused active gastritis in germ-free mice. *L. salivarius* given after *H. pylori* implantation could eliminate colonization by *H. pylori.* The findings suggest the possibility of lactobacilli being used as probiotic agents against *H. pylori.*

REFERENCES

1. Bellomo, G., Mangiagle, A., Nicastro, L., and Frigerio, G. A controlled double-blind study of SF68 strain as a new biological preparation for the treatment of diarrhea in pediatrics. *Curr. Ther. Res.* **28**(6): 927–936, 1980.
2. Biller, J. A., Katz, A. J., Flores, A. F., Buie, T. M., and Gorbach, S. L. Treatment of recurrent *Clostridium difficile* colitis with *Lactobacillus* GG. *J. Pediatr. Gastroenterol. Nutr.* **21:** 224–226, 1995.
3. Borgia, M., Sepe, N., Brancato, V., and Borgia, R. A controlled clinical study on *Streptococcus faecium* preparation for the prevention of side reactions during long-term antibiotic treatments. *Curr. Ther. Res.* **31**(2): 265–271, 1982.
4. Boudraa, G., Touhami, M., Pochart, P., Soltana, R., Mary, J.Y., and Desjeux, J.F. Effect of feeding yogurt versus milk in children with persistent diarrhea. *J. Pediatr. Gastroenterol. Nutr.* **11:** 509–512, 1990.
5. Gorbach, S. L., Chang, T. W., and Goldin, B. Successful treatment of relapsing *Clostridium difficile* colitis with *Lactobacillus* GG. *Lancet* **26:** 1519, 1987.
6. Gotz, V., Romankiewicz, J. A., Moss, J., and Murray, H. W. Prophylaxis against ampicillin-associated diarrhea with a *Lactobacillus* preparation. *Am. J. Hosp. Pharm.* **36:** 754–757, 1979.
7. Hitchins, A. D., Wells, P., McDonough, F. E., and Wong, N. P. Amelioration of the adverse effect of a gastrointestinal challenge with *Salmonella enteritidis* on weaning rats by a yogurt diet. *Am. J. Clin. Nutr.* **41:** 92–100, 1985.
8. Hotta, M., Sato, Y., Iwata, S., Yamashita, N., Sunakawa, K., Oikawa, T., Tanaka, R., Watanabe, K., Takayama, H., Yajima, M., Sekiguchi, S., Arai, S., Sakurai, T., and Mutai, M. Clinical effects of *Bifidobacterium* preparations on pediatric intractable diarrhea. *Keio J. Med.* **36:** 298–314, 1987.
9. Isolauri, E., Juntunen, M., Rautanen, T., Sillanaukee, P., and Koivula, T. A human *Lactobacillus* strain (*Lactobacillus casei* sp. strain GG) promotes recovery from acute diarrhea in children. *Pediatrics* **88**(1): 90–97, 1991.

10. Kabir, A. M. A,, Aiba, Y., Takagi, A., Kamiya, S., Miwi, T., and Koga, Y. Prevention of *Helicobacter pylori* infection by lactobacilli in a gnotobiotic murine model. *Gut* **41:** 49–55, 1997.
11. Majamaa, H., Isolauri, E., Saxelin, M., and Vesikari, T. Lactic acid bacteria in the treatment of acute rotavirus gastroenteritis. *J. Pediatr. Gastroenterol. Nutr.* **20**: 333–338, 1995.
12. Millar, M. R., Bacon, C., Smith, S. L., Walker, V., and Hall, M. A. Enteral feeding of premature infants with Lactobacillus GG. *Arch. Dis. Child.* **69:** 483–487, 1993.
13. Nomoto, K., Nagaoka, M., Yokokura, T., and Mutai, M. Augmentation of resistance of mice to bacterial infection by a polysaccharide-peptidoglycan complex (PSPG) extracted from *Lactobacillus casei. Biotherapy* **1:** 169–177, 1989.
14. Nomoto, K., Yokokura, T., Tsuneoka, K., and Shiita, M. Radioprotection of mice by a single subcutaneous injection of heat-killed *Lactobacillus casei* after irradiation. *Radiat. Res.* **125:** 293–297, 1991.
15. Oksanen, P. J., Salminen, S., Saxelin, M., Hamalainen, P., Arja, I. V., Leena, M. I., Seppo, N., Oksanen,T., Porsti, I., Salminen, E., Siitonen, S., Stuckey, H., Topilla, A., and Vapaatalo, H. Prevention of traveler's diarrhea by *Lactobacillus* GG. *Ann. Med.* **22:** 53–56, 1990.
16. Raza, S., Graham, S. M., Allen, S. J., Sultana, S., Cuevas, L., and Hart, C. A. *Lactobacillus* GG promotes recovery from acute nonbloody diarrhea in Pakistan. *Pediatr. Infect. Dis. J.* **14:** 107–111, 1995.
17. Roach, S., and Tannock, G. W. Indigenous bacteria that influence the number of *Salmonella typhimurium* in the spleen of intravenously challenged mice. *Can. J. Microbiol.* **26:** 408–411, 1980.
18. Salminen E. I., Minkkinen, J., Vapaatalo, H., and Salminen, S. Preservation of intestinal integrity during radiotherapy using live *Lactobacillus acidophilus* cultures. *Radiology* **39:** 435–437, 1988.
19. Tsuneoka, K., Ishihara, H., Dimchev, A. B., Nomoto, K., Yokokura, T., and Shikita, M. Timing in administration of a heat-killed *Lactobacillus casei* preparation for radioprotection in mice. *J. Radiat. Res.* **35**(3): 147–156, 1994.

4.1.2 Nutritional Effects

Effects	Probiotics	Test Animal	Ref.
Produce water-soluble vitamins: thiamine, nicotinic acid, folic acid, pyridoxin, vit. B_{12}	*B. bifidum* E-319 *B. infantis* S-12, 659 *B. breve* S-1, A_s-50 *B. longum* E-1946	—	1
Produce biotin	*B. adolescentis* M101-4 *B. bifidum* A234-4 *B. breve* I-53-8 *B. infantis* I-10-5 *B. longum* M101-2	—	3
Increase bioavailability of iron	*L. acidophilus* SBT 2062	Wistar rats	4

(continued)

Effects	Probiotics	Test Animal	Ref.
Deconjugation of taurocholic acid and taurodeoxycholic acid (bile salts)	*L. reuteri* 100-23 *L. delbrueckii* 100-18 *L. fermentum* 100-20 *L. delbrueckii* 100-21	LF BALB/c mice	6
Partial digestion of lactose in milk	Yogurt	Healthy human	2, 5

Production of Vitamins by Bifidobacteria

Production of Water-Soluble Vitamins (1)

Species	Strains	Sources[a]	Thiamine[b, c] (μg/ml)	Nicotinic acid[b, d] (μg/ml)	Folic acid[b,e] (μg/ml)	Pyridoxine[b, f] (μg/ml)	Vit. B_{12}[b, g] (μg/ml)
B. bifidum	E-319 Ga-18 Xa-31 I-2	Adult[h] Bottle-fed infant[i] Bottle-fed infant[i] Adult[i]	0.24 ± 0.03	1.06 ± 0.15	0.06 ± 0.02	0.04 ± 0.02	0.64 ± 0.08
B. infantis	S-12 659 S-76e La-10 Ua-60	Infant[h] Infant[h] Infant[h] Breast-fed infant[i] Breast-fed infant[i]	0.21 ± 0.04	1.22 ± 0.22	0.04 ± 0.01	0.05 ± 0.01	0.37± 0.20
B. breve	As-50 S-1 Ka-2 Va-SN Na-5	Infant[h] Infant[h] Breast-fed infant[i] Breast-fed infant[i] Breast-fed infant[i]	0.09 ± 0.08	0.39 ± 0.53	0.008 ± 0.008	0.02 ± 0.01	0.47 ± 0.42
B. adolescentis	E-194a Go-28 SI-30 YJ-9 SB-19	Adult[h] Infant[i] Adult[i] Adult[i] Breast-fed infant[i]	0.02 ± 0.02	0.18 ± 0.34	0.01 ± 0.01	0.04 ± 0.04	0.33 ± 0.21
B. longum	E-194b 5a-15 I-3 TM-25 5C-I	Adult[i] Breast-fed infant[i] Adult[i] Adult[i] Bottle-fed infant[i]	0.09 ± 0.07	0.58 ± 0.19	0.02 ± 0.01	0.04 ± 0.02	0.42 ± 0.26

[a]Derived from feces of the indicated human origin.

(continued)

[b]Figures are average value obtained with the several test straims (± SD). For vitamin determinations, the organisms were cultured for 48 h at 37°C in semisynthetic medium (SSM) containing (per liter): Proteoliquifase-digested skim milk (500 ml), $(NH_4)_2SO_4$ (10 g), KH_2PO_4 (6 g), K_2HPO_4 (8 g), $MgSO_4 \cdot 7H_2O$ (0.25 g), $FeSO_4 \cdot 7H_2O$ (12 mg), $MnSO_4 \cdot 7H_2O$ (8 mg), pyruvic acid (0.2 g), L-cysteine-HCl (1 g), riboflavin (20 mg), calcium pantothenate (2 mg), biotin (0.02 mg), adenine (20 mg), uracil (20 mg), Tween 80 (2 g), Na_2CO_3 (8 g), resazurine (2 mg), D-lactose (40 g).
[c]Thiamine was determined by the thiochrome method.
[d]Nicotinic acid was determined by microbiological assay method using *L. plantarum* ATCC 8014.
[e]Folic acid was determined by microbiological assay method using *S. faecalis* ATCC 8043.
[f]Pyridoxine was determined by microbiological assay method using *Saccharomyces carlesbergenesis* ATCC 9080.
[g]Vitamin B_{12} was determined by microbiological assay method using *L. leichmanii* ATCC 7830.
[h]Obtained from Dr. T. Mitsuoka, Department of Biomedical Science, Faculty of Agriculture, University of Tokyo, Japan.
[i]Isolated by the authors.

Remarks Many strains of the bifidobacteria investigated could synthesize thiamine, nicotinic acid, folic acid, pyridoxine, and vitamin B_{12}, except for riboflavin. *B. bifidum* and *B. infantis* could synthesize thiamine, nicotinic acid, and folic acid in significantly higher concentration while many strains of *B. breve* and *B. longum* accumulated them in lower concentration. On the other hand, none of these three vitamins were detected in most strains of *B. adolescentis*.

Production of Biotin (3)

	Extracellular Biotin Production (μg/l)[b, c, d]				
Carbon Source[a]	*B. bifidum*	*B. adolescentis*	*B. longum*	*B. infantis*	*B. breve*
D-glucose	80.6	ND	ND	ND	ND
D-mannose	20.3	ND	ND	ND	ND
Lactulose	50.4	ND	ND	ND	ND
Methycellulose	31.9	ND	ND	ND	ND
Maltotetraose	16.0	7.4	23.9	13.5	23.8
Isomalto-oligosaccharide	158.0	27.4	NL	NL	NL
Fructo-oligosaccharide	10.8	NL	43.4	15.4	NL
Galacto-oligosaccharide	9.6	37.0	3.8	13.8	36.9

[a]Bifidobacteria strains were grown in yeast extract medium containing 2% carbon source, 1.5% polypepton, 0.5% yeast extract, 0.5% NaCl, 0.03% L-cystein·HCl, in tap water, adjusted to pH 7.2, and incubated at 28°C for 7 days under anaerobic conditions.
[b]Biotin in the culture fluid was determined by a microbiological method using *Saccharomyces cerevisiae*.
[c]*B. bifidum* A234-4, *B. adolescentis* M101-4, *B. longum* M101-2, *B. infantis* I-10-5, and *B. breve* I-53-8 were obtained from Japan Bifidus Foundation.
[d]ND, not done; NL, negligible.

Remarks *B. bifidum* grew well on yeast extract medium containing oligosaccharide, isomalto-oligosaccharide, and produced the most biotin extracellularly.

Increase Bioavailability of Iron (4)

Experimental Protocol

1. Twenty-four female Wistar rats at 3 weeks old were purchased from Charles River Japan Inc. (Ibaraki, Japan). They were fed an iron-deficient diet that contained 4.7 mg/kg iron for 13 days to induce hemoglobin depletion.
2. Rats were then assigned to one of four groups (n = 6/group), such that the average initial body weight were similar among the groups.
3. For hemoglobin regeneration, they were fed ferrous-sulfate-supplemented diets. Among the two groups fed the same diet, one received oral administration of 2 ml skim milk twice a day for 7 days, the other received skim milk fermented by *L. acidophilus* SBT 2062 (cultured at 37°C, pH 3.8 for 16 h) in the same manner.
4. The blood hemoglobin value was determined by colorimetric method with Hemoglobin B Test Wako (Wako Pure Chemical Industries, Tokyo). Iron intake was calculated from food intake and dietary iron concentration. Hemoglobin regeneration efficiency was calculated from the initial and final hemoglobin values, the body weights, and the iron intake.

Dietary Iron (mg/kg)	Test Diet	Initial Hemoglobin (g/l)	Final Hemoglobin (g/l)	Iron Intake (μg/day)	HRE[a] (%)
13.7	SM[a]	77 ± 9	93 ± 9	186 ± 6	73 ± 5
13.7	FSM[a]	78 ± 8	98 ± 9	188 ± 17	84 ± 9
21.7	SM	78 ± 8	118 ± 8	316 ± 35	80 ± 6
21.7	FSM	78 ± 6	124 ± 8	326 ± 20	95 ± 10

[a]SM, skim milk; FSM, fermented skim milk; HRE, hemoglobin regeneration efficiency.

Remarks The hemoglobin regeneration efficiency was significantly higher in the group fed with fermented milk than the skim milk groups ($p \leq 0.01$). The results suggested that *L. acidophilus* SBT 2062 was effective for increasing the bioavailability of iron in rats.

Deconjugation of Bile Salts (6)

Mouse Group[a]	No. of Lactobacilli[b] ($\log_{10}$/g organ ± SE)		No. of Enterococci[b] ($\log_{10}$/g organ ± SE)		Hydrolase Activity[c] (nM cholic acid released/ 30 min/g content)	
	Ileum	Cecum	Ileum	Cecum	Ileum	Cecum
LF	—	—	—	—	60	760
LF + lactobacilli (strains 100-23, 100-21)	8.2 ± 0.1	8.3 ± 0.1	—	—	3490	5130

(continued)

Mouse Group[a]	No. of Lactobacilli[b] (log_{10}/g organ ± SE)		No. of Enterococci[b] (log_{10}/g organ ± SE)		Hydrolase Activity[c] (nM cholic acid released/ 30 min/g content)	
	Ileum	Cecum	Ileum	Cecum	Ileum	Cecum
LF + enterococci	—	—	4.5 ± 0.1	6.2 ± 0.2	180	980
RLF	—	—	4.3 ± 0.2	4.8 ± 0.2	540	3120
RLF + lactobacilli (strains 100-18, 100-20, 100-21)	7.6 ± 0.2	7.7 ± 0.2	4.2 ± 0.4	4.2 ± 0.6	1990	6180
Conventional	8.0 ± 0.2	8.4 ± 0.2	3.4 ± 0.3	4.0 ± 0.3	3970	11960

[a]LF (lactobacillus-free) BALB/c mice harbored a complex intestinal microflora but lactobacilli and enterococci were absent. Reconstituted LF (RLF) mice were derived from LF mice by inoculation with specific bacterial cultures, harvested noncultivable microbes, and cecal homogenates from chloramphenicol-treated conventional mice. *Lactobacillus* strains were isolated from the stomachs of conventional mice and identified as *L. reuteri* (strain 100-23), *L. delbrueckii* (strains 100-18, 100-21) and *L. fermentum* (strain 100-20). Enterococci were isolated from cecal samples. *Enterococcus faecalis* and *E. faecium* were the predominant species ($n = 5$ mice/group).

[b]Enumeration of the bacteria was accomplished by homogenizing the particular section of the intestinal samples in sterile distilled water to give a 10-fold dilution. Lactobacilli were enumerated on medium 10 agar plate incubated anaerobically. Enterococci were enumerated on aerobically incubated methylene blue agar plates.

[c]Intestinal contents were homogenized in sterile water. The homogenates were lyophilized for 24 h; 50 mg lyophilized homogenates were suspended in 1 ml acetate buffer (5 mM, pH 5.6) containing 0.336% EDTA, 0.156% 2-mercaptoethanol, and 0.25% Triton X-100, inside an anaerobic box containing 10% H_2–5% CO_2 in nitrogen. The cells in the intestinal homogenate were than permeabilized with Triton X-100 and a freeze–thaw cycle in a dry-ice-acetone bath. Bile salt hydrolase activity in lumenal contents was measured radiochemically by quantitating the amount of [carboxy-^{14}C] cholic acid hydrolyzed from tauro [carbonyl-^{14}C] cholic acid. Test extract (10–15 mg) was added to a 2.0 mM reaction mixture containing taurocholic acid and sufficient [^{14}C] taurocholic acid, and incubated at 37°C for 30 min. All reactions were terminated by lowering the pH to 2.0 with 6 N HCl. Ethyl acetate (2 ml) was added to partition the [^{14}C] cholic acid into the organic phase, and 1 ml samples were removed and added to 10 ml of Aguasol-II (Du Pont Co., Wilmington, DE.) in glass scintilation vials. The radioactivity was measured and counts were corrected by using an external-standard-channels ratio and a ^{14}C quench curve.

Remarks Bile salt hydrolase activity in the ileal contents of LF mice was reduced by 86% in the absence of lactobacilli (RLF) and by greater than 98% in the absence of lactobacilli and enterococci compared with samples from conventional mice. Bile salt hydrolase activity in ilea and ceca of RLF mice colonized by lactobacilli was similar to that in samples from conventional mice. It is concluded that some indigenous lactobacilli are the main contributors to total bile salt hydrolase activity in murine intestinal tract. Bile salt hydrolase-producing bacteria have been suggested to contribute to "growth depression" in production animals, a condition that can be alleviated by administration of subtherapeutic concentrations of feed additive antibiotics.

Partial Digestion of Lactose and Protein in Milk (2)

Change in Breath Hydrogen, 5 h after Ingestion of Lactose, Milk, Yogurt or Lactulose[a]

	Milk (18 g lactose in 400 ml)	Lactose (20 g in 400 ml water)	Lactulose (10 g in 200 ml water)	Yogurt (18 g lactose in 440 g)	Yogurt (11 g lactose in 270 g)
Δ Breath H (ppm)[b]	56.4 ± 3.8	56.2 ± 5.8	53.2 ± 5.3	18.1 ± 6.4	13.7 ± 4.6

[a]Values represent means ± 1 SEM for 10 subjects (healthy lactose-intolerant human on the basis of breath hydrogen concentration over 20 ppm after ingestion of 20 g lactose, 20–28 years of age).
[b]End-alveolar breath samples obtained by having the subject expired end-alveolar air into syringes fitted with stopcocks. The hydrogen concentration was analyzed by a reduction analyzer (Trace Analytical, Menlo Park, Calif.). The partial pressure of CO_2 was also determined. The observed hydrogen values were then corrected to atmospheric contamination of the alveolar air by normalization to a partial pressure of CO_2 of 45 mm Hg.

Lactase Activity in Yogurt

	Yogurt (37°C, pH 7.0), Sonicated	Yogurt (4°C, pH 4.6), Unsonicated
Δ Galactose (g/l)[a]	3.34	0.15
Lactase activity (U/g)[b]	25.0	1.1

[a]Appearance of galactose over a 4-h period was measured by a commercial kit (Boehringer Mannheim Biochemicals).
[b]One unit of lactose activity equals 1 μmol galactose produced per hour.

Remarks There is essentially no lactase activity in refrigerated yogurt at its natural pH of 4.6; the activity at a physiologic temperature and pH is appreciable.

Lactase Activity in Duodenal Contents; 20 min after Ingestion of Yogurt

	Subject 1	Subject 2	Subject 3
Lactase activity[a] (U/g duodenal juice)	8.7	10.1	12.0

[a]Subjects were intubated with a double-lumen pancreatic drainage tube. When the distal opening was in the duodenum, the subjects ingested 350 g yogurt; duodenal (and gastric) samples were obtained after 20 min. Lactase activity was determined in 1:4 dilutions of the samples in a 0.3 M phosphate buffer (pH 7.0), using a lactose concentration of 4 g per deciliter and incubation temperature of 37°C.

Remarks Lactase-deficient subjects absorbed lactose in yogurt better than lactose in milk, resulting in one third as much hydrogen excretion and fewer reports of diarrhea or flatulence. The enhanced absorption of lactose in yogurt was a result of intraintestinal digestion of lactose by lactase released from the yogurt organisms. This autodigesting feature makes yogurt a well-tolerated source of milk for lactose-deficient persons.

Percentage Lactose, Cell Counts, and Lactase Activity in the Five Dairy Products Fed to Lactase-deficient Subjects and Their Breath Hydrogen (5)

Dairy Product[a]	Percent Lactose	No. Cells per Gram Product	Lactase Activity[b] (mg/h/g product)	Change in Breath H_2[c] (ppm/h) mean ± SEM
Yogurt (500 g)	4.0	3.0×10^8	0.64	23.1 ± 16.5
Pasteurized yogurt (500 g)	4.0	3.4×10^6	0.07	122.7 ± 27.5
Cultured milk (465 g)	4.3	2.8×10^6	0.02	104.5 ± 17.1
Sweet acidophilus milk (420 g)	4.8	1.1×10^7	0	174.9 ± 27.0
Milk (pasteurized, 410 g)	4.9	Not measured	0	148.5 ± 33.0

[a]Fresh milk from the University of Minnesota dairy herd was pasteurized at 82°C for 30 min and homogenized at 2500 psi pressure. Yogurt was produced by inoculating the pasteurized milk with LBST yogurt starter culture (Marschall Products, Madison, WI), incubated at 37°C for 11 h, and cooled to 10–16°C. One half of the batch of yogurt was pasteurized at 63°C for 30 min. Another portion of the pasteurized milk was inoculated with FR1 buttermilk sour cream starter culture (Marschall Products, Madison, WI), incubated at 22°C for 16 h, then cooled to 10–16°C. Sweet acidophilus milk was produced by inoculating the pasteurized milk cooled to 7°C with *L. acidophilus* NCFM strain (GP Gunlock Co. Cincinnati, OH). The study group consisted of nine healthy subjects (20–28 y of age) who were identified as lactase deficient on the basis of a rise in breath H_2 to over 20 ppm after ingestion of milk containing 20 g lactose.

[b]Lactase activity was assayed by measurement of the initial rate of galactose appearance (Boehringer Mannheim Biochemicals, Germany) in sonicated products at 37°C.

[c]End-alveolar breath samples were obtained by having the subject expire end-alveolar air into syringes fitted with stopcocks. The hydrogen content was analyzed with a Microlyzer model 12 hydrogen gas analyzer (Quintron Instruments, Milwaukee, WI). The percentage of CO_2 in each sample was determined using a Beckman Medical Gas Analyzer (model LB-II, Beckman Instruments, Fullerton, CA). The observed H_2 values were corrected for atmospheric contamination of the alveolar air by normalization to 5% CO_2.

Remarks The degree of malabsorption of lactose (breath hydrogen production) in lactase-deficient subjects was found to correlate with the lactase activity and lactic acid bacteria number in the products. Unpasteurized yogurt is unique among the products tested in enhancing the digestion of lactose.

REFERENCES

1. Deguchi, Y., Morishita, T., and Mutai, M. Comparative studies on synthesis of water-soluble vitamins among human species of *Bifidobacteria*. *Agric. Biol. Chem.* **49**(1): 13–19, 1985.
2. Kolars, J. C., Levitt, M. D., Aouji, M., and Savalano, D. A. Yogurt—An autodigesting source of lactase. *N. Engl. J. Med.* **310:** 1–3, 1984.
3. Noda ,H., Akasaka, N., and Ohsugi, M. Biotin production by Bifidobacteria. *J. Nutr. Sci. Vitaminol.* **40:** 181–188, 1994.
4. Oda, T., Kado-oka, Y., and Hashiba, H. Effect of *Lactobacillus acidophilus* on iron bioavailability in rats. *J. Nutr. Sci. Vitaminol.* **40:** 613–616, 1994.

5. Savaiano, D. A., AbouElAnouar, A., David, E S., and Levitt, M. D. Lactose malabsorption from yogurt, pasteurized yogurt, sweet acidophilus milk, and cultured milk in lactase-deficient individuals. *Am. J. Clin. Nutr.* **40:** 1219–1223, 1984.
6. Tannock, G. W., Dashkevicz, M. P., and Feighner, S. D. Lactobacilli and bile salt hydrolase in the murine intestinal tract. *Appl. Environ. Microbiol.* **55:** 1848–1851, 1989.

4.1.3 Treatment and Prevention of Constipation

Effect of* B. breve *Strain Yakult (BBG) with Transgalactosyl Oligosaccharide (TOS) (5)

Disease[a]	No. of Cases[b]	No. of Effective Cases[c]	Efficiency (%)
Cerebrovascular diseases	6	3 }	63
Disorders of the spinal cord	2	2 }	
Diabetes mellitus, cholelithiasis	8	5	63
Hepatitis, cirrhosis of the liver	4	2	50
Intraperitoneal adhesion	4	2	50
Sigma elongation	3	2 }	63
Irritable bowel syndrome	5	3 }	
Miscellaneous			
Multiple gastric ulcer	1	0	
Chronic hepatitis	1	0	
Total	34	19	56

[a]Various diseases that result in chronic constipation in patients.
[b]Constipated patients received oral dose of BBG (6 g/day, 1–3 × 10^9/g) together with TOS (15 g/day).
[c]No. of patients in which symptoms of chronic constipation improved.

Remarks Administration of BBG (*B. breve* strain Yakult) with TOS on constipated patients showed improved symptoms in 19 out of 34 patients (56% efficiency).

	Improved Cases[b] (n = 10)		Noneffective or Slightly Improved Cases[c] (n = 8)	
Organisms[a]	Before	After	Before	After
Total flora	10.36	10.58	10.49	10.48
Bifidobacterium	9.18	9.52	8.98	9.38
Bacteroidaceae	10.03	10.16	10.12	9.91

[a]No. of organisms in intestinal flora. Units is log N/g.
[b]Ten cases were studied, and thereafter the average of the bacterial number were taken. These cases were constipated patients who showed significant improvements after administration of BBG with TOS.
[c]Eight cases were studied, and thereafter the average of the bacterial numbers were taken. These cases were constipated patients who showed noneffective or slight improvements after administration of BBG with TOS.

Remarks Administration of BBG with TOS on constipated patients showed increase in total bacterial count and bifidobacteria numbers in improved cases. In noneffective or slightly improved cases, only bifidobacteria numbers increased.

Effect of L. rhamnosus *Strain GG (3)*

	Pretreatment[a] (n = 6)	GG treatment[b] (n = 6)	Posttreatment[c] (n = 6)
L. rhamnosus GG (CFU/g feces)	<10^3–10^3	10^7	<10^3–10^5
Enzyme activities (nmol/min/mg protein)[d]			
β-glucuronidase[e]	6.0 ± 2.0	4.7 ± 1.6	5.0 ± 2.4
β-glucosidase[e]	12.1 ± 4.6	11.1 ± 5.9	11.3 ± 5.0
Glycocholic acid hydrolase[f]	15.5 ± 7.9	10.2 ± 5.4	16.7 ± 9.0
Urease[g]	75.4 ± 44.6	46.1 ± 38.8	68.8 ± 46.9
Tryptic activity (mg/kg feces)[h]	204 ± 252	132 ± 189	155 ± 188
Fecal pH[i]	7.1 ± 0.7	6.6 ± 0.9	6.9 ± 0.4
Weight of feces (g/day)	90 ± 25	79 ± 16	69 ± 19
Fecal quality[j]			
Soft (%)	21	12	7
Normal (%)	38	55	70
Hard (%)	41	33	23
Fecal frequency (times/week)	5.7 ± 2.0	5.5 ± 2.3	5.0 ± 1.4

Note: Each value in table represents the mean ± SD. n = number of cases studied.

[a]The study was divided into three 2-week periods. The pretreatment period was first 2 weeks during which the subjects had a glass (100-ml) of apple–peach drink twice a day.

[b]This was the next 2-week period during which the apple–peach drink was replaced by the same quantity of *Lactobacillus* GG-fermented apricot–peach whey drink. The composition of whey drink was (per 100 ml): energy 240 KJ, protein 1.4 g, fat 0.1 g, carbohydrate 12.5 g, and lactose <1.0 g. The *Lactobacillus* GG content was 10^8 CFU *Lactobacillus* GG/ml.

[c]This was the last 2-week period where the apple–peach drink was served again.

[d]Enzyme activities were done on all stools collected during the second week of each 2-week period. The fecal samples collected were frozen, weighed, and kept at –21°C until analysis for the different fecal enzyme activities. The subjects of the study were six elderly nursing home residents.

[e]Fecal β-glucuronidase and β-glucosidase activities were determined using the method described by Freeman (2).

[f]Glycocholic acid hydrolase activity was assayed at 37°C (pH 5.8) in 1.0 ml of 0.02 mol/l potassium phosphate buffer, 1 mmol/l glycocholic acid, and 0.1 ml fecal supernatant. Enzyme activity was terminated by the addition of 20% trichloroacetic acid (1.0 ml) and the suspension was centrifuged. The supernatant fluid (0.3 ml) was assayed for glycine spectrophotometricaly with the aid of ninhydrin. The estimation of hydrolase activity also included a measurement of fecal amino acids in the blank tubes. The reaction was linear for 15 min.

[g]Fecal urease activity was assayed at 37°C (pH 7.4) in 1.0 ml of 0.02 mol/l potassium phosphate buffer, 10 mmol/l urea, and 0.2 ml fecal extract. The reaction was stopped by addition of 0.2 N sulfuric acid (9.0 ml). Ammonia was determined using a specific electrode after 10 N sodium hydroxide (1.0 ml) has been added. The estimation of urease activity also included a measurement of fecal ammonia in the blank tubes. The reaction was linear for 30 min.

[h]Fecal tryptic activity was measured as described by Midtvedt et al. (4).

[i]Fecal pH value was recorded by nurses.

[j]Hardness of the stools was assessed by the subjects themselves.

Remarks Administration of the fermented whey drink resulted in colonization of feces by *Lactobacillus* GG in all the subjects. The effect of *Lactobacillus* GG treatment on bowel function was less evident. Although the consistency of the stools appeared to normalize, no significant changes in the fecal frequency, weight, and pH were observed. Results indicate that a *Lactobacillus* GG-fermented whey drink could change bacterial metabolism. The lowered enzymatic activities after *Lactobacillus* GG administration suggests that oral *Lactobacillus* GG intake may alter the production of toxic compounds in the colon by some bacterial species that predominate as a result of accumulation of fecal content in cases of constipation.

Effect of Lactulose Syrup (1)

	Treatment Period			Posttreatment Period[a]	
	Total[b]	Success[c]	Failure[d]	Need for Laxatives	No Need for Laxatives
Lactulose[e]	54	47	7	31	23
Glucose[e]	49	30	19	21	28
Significance probability		$p < 0.02$		$p = 0.20$	

[a]Posttreatment period was included in the experimental scheme in the first place to access a possible carryover effect of lactulose, which would show itself in a reduction of the number of those using laxatives in the posttreatment period. Laxatives, however, were prescribed only if defecation did not occur for more than 48 hours.

[b]Included in this study were 103 elderly patients of either sex who were regularly taking laxatives for treatment of chronic constipation. Patients were allocated at random to two groups to be treated for 3 weeks with either 50% lactulose syrup or with placebo (50% glucose syrup).

[c]Criterion by which the effectiveness of the treatment of lactulose was estimated was the need for additional laxatives during the treatment period. The treatment was considered to be a success if the patient needed no laxatives at all or only once in 21 days.

[d]Treatment was considered to be a failure if patient needed laxative more than once in 21 days.

[e]After a pretreatment period of 2 weeks, during which the frequency of defecation and the quantity and brand of laxatives were recorded daily, treatment with syrup was started after the first defecation in the third week. The initial dose of 15 ml was administered daily. The daily dose was reduced by half (8 ml daily) after 3 consecutive days with defecation, but, if no defecation occurred for more than 48 hours, the dose was doubled. If no defecation occurred on 3 consecutive days with the doubled dose, a laxative was administered according to the previous therapeutic regimes. If defecation occurred on 3 consecutive days with the doubled dose, the patient was treated again with the former dose, but, if the response on the doubled dose remained unsatisfactory, treatment with 30 ml syrup daily was continued until 3 weeks were completed.

Remarks During the treatment period, administration with lactulose to constipated patients resulted in a higher success rate than administration with placebo. As for the posttreatment period, the percentage of lactulose-treated patients who did not need laxatives was not statistically different from the percentage in the placebo group. It was therefore concluded that lactulose did not show a carryover effect.

	Total[a]	Success[b]	Failure[c]
Lactulose	31	25	6
Glucose	21	7	14
Significance probability		$p < 0.01$	

[a]Out of the 103 patients used in the treatment and posttreatment periods mentioned previously, only 52 patients who needed laxatives during the posttreatment periods were used in this study as only this group can be considered as genuinely suffering from constipation.
[b]Success and failure as previously defined.
[c]Glucose and lactulose were administered as previously mentioned for the treatment period.

Remarks Administration of lactulose in this group of truly constipated patients was significantly better than glucose in promoting defecation in constipated patients. Lactulose, a disaccharide, is not digested in the small intestine and passes unchanged into the colon where it serves as an energy source for carbohydrate splitting bacteria, predominantly *L. acidophilus* and *B. bifidus.* During this fermentation process, low molecular organic acids are formed that in turn affect colonic motility. Thus lactulose is useful in the treatment of constipation because it exerts its effect by alteration of the intestinal flora. Lactulose is a harmless substance: in recommended doses, an easy defecation is obtained without cramps. It does not affect blood sugar levels in diabetic patients and is not habit forming. The only side effect is transient gas formation and intestinal bloating.

REFERENCES

1. Casparis, A. W., Braadbaart, S., van der Bergh-Bohlken, G. E., and Mimica, M. Treatment of chronic constipation with lactulose syrup: Results of a double-blind study. *Gut* **9:** 84–86, 1968.
2. Freeman, H. J. Effects of differing purified cellulose, pectin and hemicellulose fiber on fecal enzyme in 1,2-dimethyhydrazine-induced rat colon carcinogenesis. *Cancer Res.* **46:** 5529–5532, 1986.
3. Ling, W. H., Hänninen, O., Mykkänen, H., Heikura, M., Salminen, S., and von Wright, A. Colonization and fecal enzyme activities after oral *Lactobacillus* GG administration in elderly nursing home residents. *Ann. Nutri. Metab.* **36:** 162–166, 1992.
4. Midtvedt, T., Carlstedt-Duke, B., Höverstad, T., Midtvedt, A. C., Norin, K. E., and Saxerholt, H. Establishment of biochemically active intestine ecosystem in ex-germfree rats. *Appl. Environ. Microbiol.* **53:** 2866–2871, 1987.
5. Shimoyama, T., Hori, S., Tamura, K., Yamamura, M., Tanaka, M., and Yamazaki, K. Microflora of patients with stool abnormality. *Bifidobacteria Microflora* **3**(1): 35–42, 1984.

4.1.4 Effect on Cholesterol Level

Subject	Probiotic	Route of Administration	Viability	Effect[a]	Ref.
Healthy human	*L. bulgaricus* *S. thermophilus*	Oral	Viable	Lower serum cholesterol level; persist for about 1 week after intake	6
Healthy human	*L. bulgaricus* *S. thermophilus*	Oral	Viable	Lower serum cholesterol level after 1 week, and maintained for the next 3 weeks after intake. So did milk and pasteurized yogurt	3
Healthy human	*Enterococcus faecium* *S. thermophilus*	Oral	Viable	Lower LDL cholesterol, no change in HDL cholesterol and triglyceride	1
Healthy human	*L. bulgaricus* *S. thermophilus*	Oral	Viable	Skim milk lowers serum cholesterol but not the yogurt	8
Healthy human	*L. acidophilus* ATCC 4962 *L. bulgaricus* ATCC 33409	Oral	Viable	No significant change	5
Healthy human	*L. bulgaricus* *S. thermophilus*	Oral	Viable	Lower serum cholesterol level after 7 days, but the effect could not be maintained by continuous yogurt consumption	4
Hypercholesterolemic rat	*Bacillus subtilis* *B. natto* *B. megaterium* *Lactobacillus acidophillus* *L. plantarum* *L. brevis* *L. casei* *Streptococcus faecalis* *S. lactis* *S. thermophilus* *Saccharomyces cerevisiae* *Candida utilis*	Oral	Viable	Lower total cholesterol level in hypercholesterolemic rats but not in normal rats. Lower low-density lipoprotein and liver cholesterol level	2
Hypercholesterolemic rat	*L. acidophilus* SBT 2062	Oral	Viable	Lower liver but not serum cholesterol level	7

[a]LDL, low-density lipoprotein; HDL, high-density lipoprotein.

Lowering Serum Cholesterol Level in Healthy Human

The Effect of Various Intakes of Milk on Cholesteremia (6)

	Serum Cholesterol in mg/dl[b]									
Trials[a]	Pre	4	8	12	16	20	24	28	32	36
A	208	200	209	186	173[c]	169[c]	172	190	214	211
B	193	199	165[c]	175[c]	169[c]	177[c]	—	196	—	198
C	211	208	196	150	162[c]	—	181	—	202	218
D	196	206	206	177	188	200	179	—	—	—

[a]Trials A: 4 liter whole milk yogurt daily for 12 days, 3 males + 1 female, 24–55 years old; B: 2 liter whole milk yogurt daily for 12 days, 4 males + 2 females, 24–55 years old; C: 2 liter skim milk yogurt daily for 12 days, 3 males + 2 females, 24–55 years old; D: 2 liter whole fresh milk daily for 12 days, 3 males + 1 female, 24–55 years old. Dannon's yogurt containing *L. bulgaricus* and *S. thermophilus* was used to ferment the milk.

[b]G. V. Mann. A method for measurement of cholesterol in blood serum. *Clin. Chem. (Winston-Salem, NC)* 7: 27, 1961.

[c]$p \leq 0.05$.

Remarks Large dietary intakes of yogurt were found to lower cholesteremia in humans. The effect appeared slowly and persisted for about 6 days after intake of the yogurt stopped, suggesting that the mechanism involved the synthesis of a regulatory protein rather than an allosteric effect. The effective agent was postulated to be hydroxymethyl glutarate.

Effect of Milk Products on Serum Cholesterol (3)

				Cholesterol (mg/day)[b]			
Study	Group	Diet	*N*[a]	Initial	1st period	2nd period	3rd period
I[c]	A	Yogurt[d], control, milk[e]	9	202 ± 32	191 ± 35	198 ± 28	207 ± 28
	B	Milk, control, yogurt	8	205 ± 30	196 ± 32	213 ± 44	189 ± 28
II[f]	C	Nonpasteurized yogurt	11	252 ± 25	230 ± 30	—	—
	D	Pasteurized yogurt	10	245 ± 24	232 ± 21	—	—
	E	Milk	10	238 ± 20	224 ± 18	—	—
	F	Control	5	244 ± 18	254 ± 11	—	—

[a]54 (24 men and 30 women) healthy volunteers, aged between 21 and 55 years.

[b]Blood sample was collected from subjects and assayed for serum cholesterol using the enzymatic method. Results expressed in mean ± SD.

[c]For group I, diet was changed accordingly every 4-week period.

[d]240 ml/day yogurt from Dannon Milk Products, Long Island City, NY, contained a live culture of *L. bulgaricus* and *S. thermophilus*.

[e]720 ml of 2% butter milk fat.

[f]For group II, diet was fed over a period of 12 weeks.

Remarks

Study I The serum cholesterol of group A fell significantly after 1 week's yogurt supplement but did not change significantly for the next 3 weeks. During the next 4 weeks of control diet, the level rose ($p > 0.05$). When diets were supplemented with milk, the serum cholesterol continued to rise. In group B, the serum cholesterol level fell after 1 week supplement and did not change during the rest of the period. During the course of the control diet, the level of cholesterol rose significantly. During the yogurt supplementation, the level fell significantly again.

Study II In groups C, D, and E, the serum choloesterol level all dropped below the initial level. Only with control group F did the level rise. Overall, the data indicate that yogurt supplementation of diet caused a significant reduction of serum cholesterol.

Effect of a New Fermented Milk Product on Lipoprotein Levels (1)

	0 weeks	3 weeks	6 weeks
Total cholesterol[a]			
Gaio[b] ($n = 29$[c])	6.08 ± 0.42	5.87 ± 0.43[d]	5.71 ± 0.49[e]
Placebo[f] ($n = 28$[c])	5.88 ± 0.47	5.85 ± 0.62	5.86 ± 0.53
HDL cholesterol[g]			
Gaio ($n = 29$)	1.21 ± 0.26	1.22 ± 0.27	1.23 ± 0.29
Placebo ($n = 28$)	1.32 ± 0.31	1.34 ± 0.30	1.31 ± 0.26
LDL cholesterol[h]			
Gaio ($n = 29$)	4.30 ± 0.34	4.08 ± 0.43[c]	3.87 ± 0.48[d]
Placebo ($n = 28$)	4.01 ± 0.51	3.98 ± 0.70	4.03 ± 0.58
Triglyceride[a]			
Gaio ($n = 29$)	1.28 ± 0.37	1.26 ± 0.53	1.35 ± 0.51
Placebo ($n = 28$)	1.22 ± 0.33	1.18 ± 0.41	1.14 ± 0.36

[a]For measurement of total cholesterol and triglyceride in plasma, enzymic methods (CHOD-PAP and GPO-PAP, Boehringer Mannheim, Germany) were used.

[b]Fermented milk product (Gaio) was produced by the research center of the Danish dairy corporation MD Foods (Aarhus, Denmark), containing 2×10^8 CFU/ml of *Enterococcus faecium* (human species) and 7×10^8 CFU/ml of two strains of *S. thermophilus*.

[c]Healthy male volunteers of Danish descent, all born in 1949. Selected on basis of having normal fasting values of plasma cholesterol (5.0–6.5 mmol/l) and *s*-triglyceride (<5 mmol/l). No history of cardiovascular, cerebrovascular, or metabolic disease, normal weight (body mass index <27.5), alcohol consumption <315 g/week. During the intervention period, the subjects' habitual diets were supplemented with 200 ml/day of either a fermented milk product or a placebo product.

[d]$p < 0.05$.

[e]$p < 0.001$.

[f]Placebo was of identical composition, but with delta-glucolactone added instead of live bacterial culture.

[g]HDL cholesterol was measured after sedimentation of apolipoprotein-B-containing lipoproteins with magnesium–phosphorous–wolfram acid (Boehringer, Mannheim, Germany).

[h]LDL cholesterol = plasma cholesterol – (HDL cholesterol + plasma triglyceride × 0.45).

Remarks After those 6 weeks, total cholesterol was reduced significantly in the group given fermented milk (–0.37 mmol/l, confidence interval: –0.51 to –0.23) while no change was observed in the placebo group (–0.02 mmol/l) ($p < 0.01$). This reduction in total cholesterol could be completely ascribed to a fall in LDL cholesterol by 10% (i.e., –0.42 mmol/l) since HDL cholesterol and triglyceride were unchanged in both groups.

No Effect on Serum Cholesterol Level in Healthy Human

Experimental Protocol (8) Of the 32 healthy adolescent school boys aged between 16 and 18, none was obese. Subjects were given 2 liters of skim milk, yogurt, or full cream milk daily for 3 weeks. Blood samples were obtained and was allowed to clot at room temperature, spun down, and the serum analyzed manually for total cholesterol (Boehringer enzymatic method), and HDL cholesterol (dextran sulfate-magnesium chloride preparation). LDL cholesterol was calculated by the formula: Total cholesterol = high density cholesterol + total glycerides/5.

	Weeks					
	Baseline		Experimental			Follow up
	I	II	1	2	3	III
Skim milk (0.2% fat)						
Total cholesterol	190 ± 34[a]	179 ± 27	172 ± 33	166 ± 34[b]	154 ± 32[c]	161 ± 28[c]
LDL cholesterol	133 ± 31	131 ± 26	123 ± 34[a]	118 ± 35[a]	117 ± 34[a]	115 ± 29[a]
HDL cholesterol	46 ± 12[a]	39 ± 6	46 ± 10[a]	40 ± 7	29 ± 8[b]	37 ± 5
Yogurt (1.8% fat)						
Total cholesterol	154 ± 32[a]	146 ± 19	156 ± 28[a]	171 ± 36[c]	141 ± 30	140 ± 28
LDL cholesterol	96 ± 20	104 ± 33	100 ± 33	116 ± 37[c]	101 ± 31	98 ± 28
HDL cholesterol	45 ± 12[a]	34 ± 10	45 ± 10[c]	41 ± 5[a]	31 ± 6	35 ± 14
Full cream milk (3.3% fat)						
Total cholesterol	185 ± 31[a]	172 ± 29	189 ± 36[c]	196 ± 22[c]	169 ± 36	172 ± 28
LDL cholesterol	133 ± 22	118 ± 26	123 ± 33	127 ± 28	117 ± 37	120 ± 26
HDL cholesterol	47 ± 10	42 ± 12	56 ± 11[a]	55 ± 12[c]	44 ± 12	46 ± 11

[a] $p < 0.05$.
[b] $p < 0.025$.
[c] $p < 0.01$ versus week 2, Wilcoxon signed-rank test.

Remarks Changes in serum total cholesterol and lipoprotein cholesterol correlated with dietary fat and cholesterol intakes. No convincing evidence of a cholesterol-lowering factor in yogurt could be found.

Experimental Protocol (5) A commercially available Lactinex preparation (Becton Dickinson Microbiology Systems, Cockeysville, MD) containing about 2×10^6 CFU/tablet of *L. acidophilus* ATCC 4962 and *L. bulgaricus* ATCC 33409 or placebo tablets were given to 354 nonfasting healthy adults in a dose of one tablet each, taken 4 times a day. There was a 3-week washout period between the two 6-week treatment periods.

Component	Time			
	First 6 weeks		Second 6 weeks	
	0	6	0	6
Placebo-treated group				
TC[a]	208	215	211.3	213.4
HDL[b]	50	53.0	53.0	56.7
LDL[c]	134.1	136.7	136.7	134.3
Lactobacillus-tablets-treated group				
TC[a]	206.3	210.0	214.9	219.3
HDL[b]	52.2	54.0	52.2	55.9
LDL[c]	131.2	132.2	139.0	139.8

[a]Average serum total cholesterol concentration, mg/dl. For measuring cholesterol, an enzymatic procedure available from Sigma Chemical Co., St. Louis, MO, was used.
[b]Average serum high-density lipoprotein concentration, mg/dl.
[c]Average serum low-density lipoprotein concentration, mg/dl.

Remarks Overall, there is no significant change in TC, LDL, and HDL levels throughout the study for the placebo-treated or the *Lactobacillus*-treated group.

Temporary Effect on Serum Cholesterol Level in Healthy Human (4)

Experimental Protocol Ten healthy adult men, aged 23–39 years, were given 681 g nonfat, unpasteurized yogurt daily. Three strains of each of the *L. bulgaricus* and *S. thermophilus* cultures were used to produce the yogurt; strains CH-I and CH-II were obtained from Chr. Hansen's Laboratory, Inc., Milwaukee, WI. The SH-III strains were supplied by Dr. K. M. Shahani, University of Nebraska, Lincoln, NE.

Total serum cholesterol was measured enzymatically using a kit (Beckman Instruments, Inc., Brea, CA). High-density lipoprotein cholesterol was measured after precipitation of low-density lipoprotein cholesterol from the serum by dextran sulfate and magnesium acetate. Serum LDL was calculated from the difference between total and HDL cholesterol.

	Diet-period (days consumed)[a]									
	CH-I, $n = 8$			CH-II, $n = 10$			SH-III, $n = 10$			
	0	7	14	0	7	14	0	7	14	21
Total cholesterol	189	167	178	182	163	165	178	173	156	158
(mg/dl)	± 14	± 9[a]	± 13	± 8	± 8[b, c]	± 7[b]	± 8	± 8[c]	± 9[b–d]	± 10[b]
HDL cholesterol	44	41	43	48	33	43	42	41	42	39
(mg/dl)	± 4	± 3	± 4	± 3	± 3[b, c]	± 3[d, e]	± 3	± 3	± 3	± 2
LDL cholesterol	121	106	115	113	113	94	118	114	95	101
(mg/dl)	± 13	± 10	± 13	± 10	± 7	± 5[c–e]	± 7	± 7	± 8[d, e]	± 10

[a]CH-I and CH-II lasted 14 days, SH-III lasted 21 days.
[b]Significantly different ($p < 0.05$) from day 0 within same diet period.

[c]Significantly different ($p < 0.05$) same day between diet periods.
[d]Significantly different ($p < 0.05$) from day 7 within same diet period.
[e]Significantly different ($p < 0.05$) from day 7 between diet periods.

Remarks Overall the results suggested that yogurt consumption by human males lowered total serum cholesterol after 7 days, but the hypocholesterolemic effect was not maintained by continued yogurt consumption. The high-density cholesterol (HDL-C) remained unchanged throughout the diet periods CH-I and SH-III, although on the seventh day of CH-II, a 30% decline was observed ($p < 0.05$). Serum LDL cholesterol generally followed the response of total serum cholesterol. On diet CH-I, LDL-C dropped significantly after 14 days ($p < 0.05$). Differences in concentrations of uric, orotic, and hydroxymethyglutaric-like acids in yogurts, which have been suggested likely hypocholesterol agents were insufficient to account for the temporary hypocholesterolemic effects of yogurt made from various strains.

Lowering of Serum and Liver Cholesterol Level in Diet-Induced Hypercholesterolemic Rat (2)

Experimental Protocol Male F344 rats (10 per group, purchased from CLEA Japan Inc., Tokyo) at 8 weeks of age were fed either a basal diet or a high-fat, high-cholesterol diet containing 10 g cholesterol/kg for 4 weeks. The control diet was supplemented with 150 g rice bran/kg, whereas the probiotic group was supplemented with 150 g probiotic/kg, for 6 weeks. The probiotic contained each of the *B. subtilis, B. natto, B. megaterium, L. acidophilus, L. plantarum, L. brevis, L. casei, S. faecalis, S. lactis, S. thermophilus, Saccharomyces cerevisiae,* and *Candida utilis,* at 10^{7-8} CFU/g rice bran. Total cholesterol and HDL cholesterol concentrations in the serum were determined enzymatically using commercially available reagent kits (assay kits for the TDX system, Abbott Lab Co. Irvine, CA, USA). The difference between total and HDL cholesterol was taken to be the concentration of VLDL, IDL, and LDL cholesterol.

	High Fat, High Cholesterol Diet		Basal Diet	
Component	Control	Probiotic	Control	Probiotic
Serum (mmol/l)				
Total cholesterol	215.0	180.7[a]	132.9	127.0
HDL cholesterol	1.1	1.3	1.7	1.9[a]
VLDL + IDL + LDL cholesterol	3.8	2.8[a]	1.1	0.5[a]
Liver (μmol/g dry liver)				
Cholesterol	105.5	51.6[a]	27.0	21.9[b]

[a]Significant difference from the controls at $p < 0.01$.
[b]Significant difference from the controls at $p < 0.05$.

Remarks There were significant differences between control and probiotic treatments of rats fed on high-fat, high-cholesterol diet. However, those rats fed on basal diet with probiotic were not significantly different from those in the control group. The dietary probiotic promoted serum HDL cholesterol concentrations but decreased the VLDL, IDL, and LDL cholesterol concentration in both diet groups. The liver cholesterol concentrations in the probiotic groups decreased significantly compared with those found in the control groups.

Lowering of Liver Cholesterol Level in Diet-Induced Hypertriglyceridemic Rats (7)

Experimental Protocol Male Sprague-Dawley rats (purchased from CLEA Japan, Inc., Tokyo) at 4 weeks of age (eight animals/group), were fed a hypertriglyceridemic diet containing 20% coconut oil, 17.5% fructose, and 17.5% sucrose for 14 days. The test diet was supplemented with either 20% skim milk powder or 20% fermented milk powder containing 1.2×10^9 CFU/g of *L. acidophilus* SBT 2062 (provided by Snow Brand Milk Products, Kawagoe, Japan). Cholesterol levels were determined by the enzymatic procedures with a Determiner TC 5 (Kyowa Medex Co., Tokyo).

Dietary Group	Plasma Cholesterol[a] (mg/100 ml)	Liver Cholesterol[a] (mg/g)
Control	78 ± 3	3.9 ± 0.1
Skim milk	78 ± 1	3.2 ± 0.1[b]
Fermented product	77 ± 3	3.1 ± 0.1[b]

[a]Values are mean ± SD for eight rats.
[b]Significant difference from the control group at $p < 0.05$ by Dunnett's *t*-test.

Remarks Both dairy products (skim milk and fermented milk) prevented the elevation of liver cholesterol level but had no effect on plasma cholesterol level.

REFERENCES

1. Agrebæk, M., Gerdes, L. U., and Richenlsen, B. Hypocholesterolaemic effect of a new fermented milk product in healthy middle-aged man. *J. Clin. Nutr.* **49:** 346–352, 1995.
2. Fukushima, M., and Nakano, M. The effect of a probiotic on faecal and liver lipid classes in rats. *Br. J. Nutr.* **73:** 701–710, 1995.
3. Hepner, G., Fried, R., St. Jeor, S., Fusetti, L., and Morin, R. Hypocholesterolemic effect of yogurt and milk. *Am. J. Clin. Nutr.* **32:** 19–24, 1979.
4. Jaspers, D. A., Massey, L. K., and Luedecke, L. O. Effect of consuming yogurts prepared with three culture strains on human serum lipoproteins. *J. Food Sci.* **49:** 1178–1181, 1984.
5. Lin, S. Y., Ayres, J. W., Winkler, W., Jr., and Sandine, W. E. Lactobacillus effects on cholesterol: In vitro and in vivo results. *J. Dairy Sci.* **72:** 2885–2899, 1989.
6. Mann, G. V. A factor in yogurt which lowers cholesteremia in man. *Atherosclerosis (Shannon, Irel.)* **26:** 335–340, 1977.
7. Oda, T., and Hashiba, H. Effects of skim milk and its fermented product by *Lactobacillus*

acidophilus on plasma and liver lipid levels in diet-induced hypertriglyceridemic rats. *J. Nutr. Sci. Vitaminol.* **40:** 617–621, 1994.

8. Rossouw, J. E., Burger, E.-M., Van Der Vyver, P., and Ferreira, J. J., The effect of skim milk, yogurt and full cream milk on human serum lipids. *Am. J. Clin. Nutr.* **34:** 351–356, 1981.

4.1.5 Treatment of Hepatic Encephalopathy (1)

Age & Sex[a]	Diagnosis	Dietary Protein (g/24 h)	Neomycin Therapy	EEG[c] (mean, c/s)		Arterial Ammonia[d] (μg/ml)	
				–Enpac	+Enpac[b]	–Enpac	+Enpac[b]
45 M	Cirrhosis, portacaval anastomosis	20	3 mo	5.6	6.2	1.44	0.93
57 F	Cirrhosis, portacaval anastomosis	40	3 mo	5.4	6.2	2.25	1.80
65 F	Portal hypertension, 2nd hepatic cancer	40	6 mo	5.8	6.6	1.28	1.15
65 M	Alcoholic cirrhosis	40	Nil		6.8	1.6	1.6
65 M	Cirrhosis	40	Nil		7.7	0.9	0.9
72 M	Cirrhosis	40	Nil		6.3		0.9
53 M	Cirrhosis, portacaval anastomosis	40	3 y	6.5	6.7	—	—
59 F	Cirrhosis	40	6 mo alt. days	5.1	6.4	—	—
38 M	Cirrhosis, portacaval anastomosis	20	6 mo	4.6	5.2	3.7	2.6
59 F	Cirrhosis	40	2 weeks	7.7	7.7	1.6	1.5

[a]Ten patients suffering from recurrent episodes of acute hepatic encephalopathy or from chronic symptoms. All diuretic therapy and sedatives were stopped for the duration of the studies. Seven of the patients received 1 g neomycin, 3 times daily.

[b]Enpac was given in a dose of 20–40 g in 4 doses daily. Enpac contained 1×10^7 *L. acidophilus*/g. Treatment was continued for 1–4 weeks. Patients also received 60 g lactose, 4 times daily.

[c]EEG rhythms were fed through an electronic waveform analyzer, slowing of which occurred in hepatic encephalopathy. Normal mean frequency is 8 or more c/s. In hepatic precoma, it is 4 c/s or less.

[d]Arterial blood was drawn from the bronchial artery of the fasting and resting patients. Normal arterial ammonia is up to 100 μg/100 ml.

Remarks In five out of seven patients with chronic hepatic encephalopathy on long-term neomycin. A freeze-dried preparation of *L. acidophilus* produced an improvement in the EEG and clinical status. A fall in blood ammonia occurred in three out of five patients tested.

REFERENCE

1. Read, A. E., McCarthy, C. F., Heaton, K. W., and Laidhaw, J. *Lactobacillus acidophilus* (Enpac) in treatment of hepatic encephalopathy. *Br. Med. J.* **1:** 1267–1269, 1966.

4.1.6 Treatment of Chronic Urinary Tract Infection (1)

	No. of *Candida*/g Feces[a]	No. of Patients	*Bifidobacterium* Administration[b]	No. of *Candida*/g Feces after *Bifidobacterium* Administration	*Candida* Infection/No. of Patients[c]
Patients receiving antileukemic therapy[d]	$\geq 10^5$	49	No	—	11/21
			Yes	$\geq 10^5$	10/14[e] (A)
				$\leq 10^4$	2/14 (B)
	$\leq 10^4$	51	No	—	3/51
Normal control	$\geq 10^5$	0	ND	—	0/34
	$\leq 10^4$	34	ND	—	34/34

[a]Isolation and enumeration of *Candida* was done on *Candida* GE culture medium (Nissui Pharmaceutical Co. Ltd., Japan). The species of *Candida* was identified by *Candida* Check (Yatron Co., Ltd., Japan). 34.7% patients were found infected by *Candida albicans* A, 16.3% by *C. albicans* A and *C. glabrata,* 18.4% by *C. glabrata,* and 16.3% by *C. albicans* A and others.
[b]Morinaga Bifidus (Morinaga Milk Industry Co. Ltd., Japan) as milk contained 10^7/ml each of *B. longum* and *L. acidophilus.* Levenin (Wakamoto Pharmaceutical Co. Ltd., Japan) contained 10^7/g *L. acidophilus, B. infantis,* and *Enterococcus faecalis.*
[c]Respiratory *Candida* infection was diagnosed when a patient with respiratory symptoms had more than 10^6 *Candida*/g sputum. Urinary infection was diagnosed when *Candida* was found in the urine of a patient with urinary symptoms.
[d]One hundred leukemia patients received ordinary doses of enocitabine, daunomycine, 6-MP, prednisolone, cyclophosphamide, busulfan, or vincristine for more than 3 months.
[e]$p < 0.01$ (significant difference, A vs. B).

Remarks Respiratory and urinary infections were higher in patients with more than 10^5 *Candida*/g feces. *Bifidobacterium* administrated orally to patients with more than 10^5 *Candida*/g feces reduced the incident of infection if the intestinal *Candida* population fell to less than 10^4/g feces.

REFERENCE

1. Tomoda, T., Nakano, Y., and Kageyama, T. Intestinal *Candida* overgrowth and *Candida* infection in patients with leukemia: Effect of Bifidobacterium administration. *Bifidobact. Microflora* **7**(2): 71–74, 1988.

4.1.7 Antitumor

Target	Subject	Probiotic	Route of Administration[a]	Viability	Effect	Ref.
Malignant pleural effusion secondary to lung cancer	Human patients with adenocarcinoma (phase III trial)	*L. casei* strain Shirota	IPL	Heat killed	Improve response to doxorubicin treatment, performance status, and symptoms. Prolong survival.	7

(*continued*)

Target	Subject	Probiotic	Route of Admini-stration[a]	Viability	Effect	Ref.
Uterine cervix carcinoma	Human patients with stage IIIB cervical cancer (phase III trial)	*L. casei* strain Shirota	Intra-dermal	Heat killed	Enhance tumor regression by radiation. Prolong survival and relapse-free interval. Prevent leukopenia during radiation therapy.	12
Enzyme catalyze procarcinogens to carcinogens	Human	*L. acidophilus* NCFM & N2	Oral	Viable	Reduce fecal bacterial β-glucuronidase, nitroreductase, azoreductase.	3
Meth A fibrosarcoma	BALB/c mice	*L. casei* strain Shirota	ipl	Heat killed	Prolong survival, as effective as OK432, *C. parvum* & BCG.	10
Meth A fibrosarcoma	BALB/c mice	*B. infantis* peptidoglycan	sc	Cell wall extract	Suppress tumor development.	14
Ehrlich ascites tumor	Swiss mice	Yogurt: *L. bulgaricus* & *S. thermophilus*	Oral	Viable	Inhibit initial tumor growth. No effect on survival rate.	13
Sarcoma 180	ICR mice	*L. casei* strain Shirota	iv	Heat killed	Suppress tumor growth.	6
MCA K-1	BALB/c	*L. casei* strain Shirota	iv and ip	Heat killed	Suppress tumor growth.	
DMH-induced intestinal tumor	Fischer 344 rat	*L. rhamnosus* GG	Oral	Viable	Suppress initiation or early promotion stages of tumorigenesis.	4
Colon 26 tumor	BALB/c mice	*L. casei* strain Shirota	Oral	Viable	Suppress secondary tumor growth.	5
Mouse bladder Tumor-2 cells	C3H/He mice	*L. casei* strain Shirota	Oral iv	Viable Heat killed	Suppress tumor growth.	1
Superficial bladder cancer	Human patient	*L. casei* strain Shirota	Oral	Viable	Prevent tumor recurrence following transurethral resection.	2
Lewis lung carcinoma	C57BL/6 mice	*L. casei* strain Shirota	ipl il/iv	Heat killed	Inhibit pulmonary metastases after inoculation of tumor.	8
			sc/il/ip	Heat killed	Inhibit growth of primary tumor and lung metastases after priming with probiotic organisms.	11

(*continued*)

Target	Subject	Probiotic	Route of Administration[a]	Viability	Effect	Ref.
B16-BL6 melanoma	C57BL/6 mice	*L. casei* strain Shirota	il/iv	Heat killed	Intralesional and injection before surgical excision of primary tumor inhibit axillay lymph node metastasis. Intravenous injection after surgical excision of tumor inhibits both axillary lymph node and lung metastases. Combination of il and iv markedly inhibit metastases.	9

[a]ipl, intrapleural; sc, subcutaneous; iv, intravenous; ip, intraperitoneal; il, intralesional.

Enhance Responses to Chemotherapy (7)

Treatment Group[a]	Number of Patients[b]	CR[c]	PR[c]	Failure[c]	Mann-Whitrey's U-Test	Chi-Square Test Response Rate (%) CP + PR
LC 9018 gp (overall)	38	12	16	10		73.7
					$p < 0.01$	$p < 0.01$
Control gp (overall)	38	7	8	23		39.5
LC 9018 gp (with adenocarcinoma)	36	11	16	9		75.0
					$p < 0.05$	$p < 0.01$
Control gp (with adenocarcinoma)	30	5	7	18		40.0

[a]The patients were randomly allocated to receive either intrapleural doxorubicin (40 mg in 20–50 ml saline) alone (control group) for 3–4 h or intrapleural doxorubicin plus a subsequent 0.2 mg of LC 9018 (heat-killed *L. casei* strain Shirota) in 20–50 ml saline (LC 9018 group) for another 3–4 h.

[b]A phase III randomized, controlled trial involved 76 patients with malignant pleural effusions secondary to lung cancer (stages III and IV). Among the patients, 66 had adenocarcinoma; 55 institutions and hospitals were involved.

[c]Complete response (CR) is negative cytologic findings with no reaccumulation of fluid for at least 4 weeks. Partial response (PR) is negative cytologic findings with asymptomatic minimal fluid accumulation (defined as an effusion that required no additional aspiration for at least 4 weeks. Failure is detectable intrapleural fluid even after tube drainage, with radiographic or computed tomographic findings indicating no improvement, or exacerbation compared with before treatment or failure to confirm conversion to negative cytologic study.

Remarks

1. The heat-killed *L. casei* strain Shirota (LC 9018) group showed significantly ($p < 0.01$) better response than the control group in patients with adenocarcinoma.
2. The overall response rate (CR + PR) was 73.7% in LC 9018 group (31.6% CR and 42.1% PR) and 39.5% in the control group (18.4% CR and 21.1% PR).
3. LC 9018 group also showed significant improvement in performance status and symptoms including anorexia, nausea, and chest pain than the control group ($p < 0.05$).

Enhance Tumor Regression by Radiation Therapy (12)

	After 30 Gy of Radiation[a]		At Completion of RT	
Tumor Response	RT-LC gp	RT gp	RT-LC gp	RT gp
CR[b]	2	1	51	45
PR[b]	50	43	37	51
MR[b]	31	48	6	7
NC[b]	12	15	1	4
CR rate (%)	2/95 (2.1)	1/107 (0.9)	51/95 (53.7)	45/107 (42.1)
CR + PR rate (%)	52/95 (54.7)	44/107 (41.1)	88/95 (92.6)	96/107 (89.7)
U test	$p = 0.085$		$p = 0.094$	

Cancer. **72**(6): 1952, 1993.

[a]A phase III, randomized, controlled study involving 228 patients with stage IIIB cervical cancer at 50 institutions. Patients were stratified according to tumor volume. They were then randomly allocated to receive either radiation therapy with LC 9018 (RT-LC group) or radiation therapy alone (total dose of 45 Gy, RT group). For RT-LC group, 0.1 mg heat-killed *L. casei* strain Shirota was administered intradermally twice a week during radiation therapy and afterward 0.1 mg/2 weeks for 2 y or until tumor recurrence. Both groups received adjuvant chemotherapy orally with 5-fluorouracil (200 mg/day) for 6 mo after radiation therapy.

[b]CR, complete response; PR, partial response, a ≥ 50% decrease in tumor size; MR, minor response, a ≥ 25% decrease in tumor size; NC, no change. Size and extent of tumor was measured by pelvic examination or imagining diagnosis.

Remarks Heat-killed *L. casei* strain Shirota (LC 9018) enhanced tumor regression ($p < 0.1$) by radiation after both 30 Gy of external radiation and at the completion of radiation therapy. The combination therapy also prolonged survival (69.2 vs. 46.2%, $p < 0.05$) and the relapse-free interval (66.5 vs. 48.5%, $p < 0.05$) over a 4-year period, compared to radiation alone. Major side effects of combined LC 9018 included fever and skin lesions at the injection site.

Suppression of Formation of Carcinogens (3)

Protocol	Days on Protocol[a]	Specific Activity (μg/min/mg protein)		
		β-glucuronidase[b]	Nitroreductase[c]	Azoreductase[d]
Baseline 1	0	1.74	6.20	4.60
	10	1.70	5.40	4.30
	20	2.14	4.30	3.90
Milk feeding	30	2.15	4.70	4.70
	40	2.00	4.00	4.30
	50	2.10	4.10	4.10
Baseline 2	60	2.20	4.50	5.10
	70	2.50	4.40	4.90
	80	2.00	4.90	5.10
Lactobacillus feeding	90	1.30	3.20	3.10
	100	1.10	1.60	2.10
	110	1.12	1.20	1.20
Baseline 3	120	1.25	2.20	2.00
	130	1.72	2.80	3.10
	140	1.90	4.20	4.10

[a]Protocol: Started with 4 weeks control period followed by 4 weeks of 500 ml/day low-fat plain milk, alternate with another 4 weeks control (no dietary supplements), 4 weeks of 500 ml/day milk containing 2 $\times 10^6$ per ml of viable *L. acidophilus* N2 or NCFM and a final 4 weeks control period. Twenty-one healthy subjects (16 women and 5 men) were tested for their fecal bacterial enzyme activity (β-glucuronidase, nitroreductase, and azoreductase). These enzymes catalyze procarcinogens to carcinogen.

[b]Assays of β-glucuronidase: (i) Fresh fecal samples (100 mg/ml) were suspended in cold 0.1 M potassium phosphate buffer (pH 7.0), blended in a prechilled Ten Broek homogenizer. The homogenate was passed through surgical gauze, the filtrate was disrupted by sonication with a Biosonik III (Bronwill, Rochester, NY) by three 1-min bursts at 4°C and then centrifuged at 500*g* for 15 min. The supernatant (fecal extract) was used for enzyme assay. (ii) The enzyme activity was determined at 37°C (pH 6.8) in 1 ml solution containing 0.02 M potassium phosphate buffer, 0.1 mM EDTA, 1 mM phenolphthalein β-D-glucuronide, and 0.1 ml fecal extract. (iii) The reaction was stopped at 5-min intervals up to 30 min by addition of 5 ml of 0.2 M glycine buffer (pH 10.4) containing 0.2 M NaCl. Readings were taken at 540 nm and plotted against time of incubation. The amount of phenolphthalein released was calculated by comparison with a standard curve. The enzyme activity was expressed as micrograms of phenolphthalein formed per hour per milligram of fecal protein.

[c]Assays of nitroreductase: (i) Fresh fecal samples (100 mg/ml) were suspended in cold 0.2 M Tris-HCl buffer (pH 7.8). The buffer was placed in an anaerobic chamber for 24 h before use. All subsequent manipulations were done under a flow of O_2-free CO_2 gas. (ii) The fecal specimens were disrupted with a spatula and agitated in a stoppered tube, containing glass beads (0.2 mm diameter), for several minutes on a vortex mixer. The supernatant was used for the enzyme assay. (iii) The assay was performed anaerobically at 37°C, in 1 ml solution containing 0.08 M Tris-HCl buffer (pH 7.8), 0.35 mM *m*-nitrobenzoic acid (Aldrich Chemical, Milwaukee, WI), 0.5 mM NADPH, 1 mM NADH, and 0.4 ml fecal extract. The reaction was stopped by addition of 1.5 ml of 1.2 N HCl within 2 h of incubation, where enzymatic reaction was linear. The amount of *m*-aminobenzoic acid formed was measured spectrophotometrically at 550 nm against known concentration of *m*-aminobenzoic acid (K & K Laboratories, Plainview, NY). The enzyme activity was expressed as micrograms of aminobenzoic acid formed per hour per milligrams of fecal protein.

[d]Assays of azoreductase: The enzyme activity was assayed anaerobically at 37°C, pH 7.8, in 1 ml solution containing 400 μm sunset yellow [FD & C yellow No. 6; 1-(1-sulfo-1-phenylozo)-2-napthol-6-sul-

fonic acid; Mallinckrodt Inc., St. Louis, MO] and 0.6 ml of fecal extract. The reaction was stopped within 2 h by addition of 1.5 ml of 1.2 N HCl. The reaction mixture was then centrifuged at 20,000*g* for 30 min, and the reduced products in the supernatant were extimated spectrophotometrically at 550 nm against a standard curve. The standard curve was constructed using various amounts of chemically reduced sunset yellow (by treatment of the dye with sodium hydrosulfite). The enzyme activity was expressed as micrograms of reduced sunset yellow per hour per milligrams of fecal protein.

Remarks Reductions of two- to fourfold in the activities of the three fecal enzymes were observed only during the period of lactobacilli feeding. These changes were noted in all subjects and were highly significant ($p < 0.02$–0.01). During the final control period, after lactobacilli feeding, fecal enzyme levels returned to normal after 4 weeks.

Suppress Primary Tumor Growth

Meth A Fibrosarcoma

Effect of LC 9018 on Survival of Meth-A-Bearing Mice (10)

Group[a]	Treatment with LC 9018	Survival Time (days, mean ± SD)	T/C[b] (%)
1	Control	7.4 ± 0.9	100
2	250 μg	9.6 ± 2.8	130
3	250 μg	8.6 ± 1.1	116
4	100 μg	10.9 ± 3.0	147
5	100 μg	10.1 ± 1.9	136

[a]Meth A cells (5×10^5/mouse) were inoculated ipl into BALB/c mice (10 mice/group) on day 0. Heat-killed *L. casei* strain Shirota (LC 9018) was injected ipl on days 1 to 5 (groups 2 and 4) or on days 1, 3, and 5 (groups 3 and 5).

[b]Survival Rate (T/C %) = $\frac{\text{mean survival days of treated group}}{\text{mean survival days of control group}} \times 100$.

Remarks The ipl treatment with heat-killed *L. casei* strain Shirota (LC 9018) significantly prolonged the survival of mice when used at a dose of 100 μg/mouse than at 250 μg/mouse.

Effect of Pretreatment with LC 9018 on Survival of Meth-A-Bearing Mice (10)

Group[a]	Treatment with LC 9018	Survival Time (days, mean ± SD)	T/C[b] (%)
1	Control	7.3 ± 0.8	100
2	250 μg	14.3 ± 3.4	196
3	250 μg	17.1 ± 6.6	234
4	100 μg	21.0 ± 3.9	288
5	100 μg	20.8 ± 5.2	285

[a]Meth A cells (5×10^5/mouse) were inoculated ipl into BALB/c mice (10 mice/group) on day 0. LC 9018 was injected ipl on days –5 to –1 (groups 2 and 4) or on days –5, –3, and –1 (groups 3 and 5).

[b]Survival Rate (T/C %) = $\frac{\text{mean survival days of treated group}}{\text{mean survival days of control group}} \times 100$.

Remarks Pretreatment with heat-killed *L. casei* strain Shirota (LC 9018) given ipl also significantly prolonged the survival of mice inoculated with Meth A ipl.

Comparison of Antitumor Effect on OK432, *C. parvum,* or BCG and LC 9018 (10)

Treatment with Adjuvants[a]	Dose (μg)	Survival Time (days, mean ± SD)	T/C[b] (%)
Control	—	10.8 ± 1.5	100
LC 9018	500	18.2 ± 4.0	169
LC 9018	250	20.0 ± 3.6	185
LC 9018	100	27.0 ± 0.7	250
OK 432	250	15.7 ± 2.2	145
OK 432	100	15.7 ± 3.6	145
C. parvum	500	11.2 ± 2.3	104
C. parvum	250	9.6 ± 1.2	89
C. parvum	100	13.0 ± 1.1	120
BCG	500	14.2 ± 3.1	131
BCG	250	13.8 ± 4.2	128
BCG	100	15.5 ± 1.1	144

[a]Meth A cells (1×10^5/mouse) were inoculated ipl in BALB/c mice (10 mice/group) on day 0. LC 9018, OK 432, *C. parvum,* or BCG was injected ipl on days 1, 3, 5, 7, and 9.

[b]Survival Rate (T/C %) = $\frac{\text{mean survival days of treated group}}{\text{mean survival days of control group}} \times 100.$

Remarks Treatment with heat-killed *L. casei* strain Shirota (LC 9018) was more effective than OK432, *C. parvum,* and BCG in prolonging the survival of mice inoculated with Meth A ipl.

In vitro Cytolytic Assay of Thoracic Exudate Cells from Mice Injected with LC 9018 ipl against Meth A Cells (10)

Days after LC 9018 Injection[a]	Cytolytic Activity		
	Whole	Adherent	Nonadherent
Untreated control	0.2 ± 1.8	−3.7 ± 1.9	1.8 ± 1.5
1	2.5 ± 1.5	1.0 ± 4.3	4.7 ± 0.2
3	32.7 ± 4.8	37.9 ± 4.4	−1.8 ± 1.6
5	28.6 ± 6.1	22.6 ± 3.7	3.3 ± 0.8
7	27.4 ± 4.6	18.9 ± 6.4	0.7 ± 2.7

[a]LC 9018 was given ipl to BALB/c mice (10 mice/group) at a dose of 100 μg/mouse. Cytolytic activity was measured by ^{51}Cr release assay. Effector/target ratio was 25/1.

Remarks The cytolytic activity of the thoracic exudate cells (TEC) against Meth A was increased (especially that of the thoracic macrophages) on day 3 and maintained for 7 days.

Tumor Suppression Test with Cell Wall Peptidoglycan (WPG) from *B. infantis* in Flora-Bearing Euthymic (+/+) and Athymic (nu/nu) Mice and Rechallenge of Tumor in Survivors (14)

Expt.	Treatment[a]	Tumor Incidence %[b]		
		+/+	nu/+	nu/nu
I	Meth A	10/10 (100)	9/10 (90)	10/10 (100)
	Meth A + WPG 100 μg	4/10 (40)[c]	4/10 (44)[d]	7/10 (70)
	Meth A + WPG 200 μg	1/10 (10)[c]	4/10 (44)[d]	5/10 (50)[d]
II	Meth A	10/10 (100)	10/10 (100)	10/10 (100)
	Meth A + WPG 100 μg	0/9 (0)[c]	1/10 (10)[c]	8/10 (80)
	Meth A + WPG 200 μg	0.9 (0)[c]	0/10 (0)[c]	6/10 (60)[d]
III	Meth A	9/10 (90)	8/10 (80)	9/9 (100)
	Meth A (survivors of WPG 100 μg in expt II)	4/9 (44)	2/9 (22)[d]	2/2 (100)
	Meth A (survivors of WPG 200 μg in expt II)	2/9 (22)[c]	3/10 (30)[d]	4/4 (100)

[a]Meth A fibrosarcoma cells (10^5) or Meth A cells (10^5) mixed with cell wall peptidoglycan extracted from *B. infantis* Reuter (ATCC 15697) in 0.1 ml PBS were injected subcutaneously along the linea alba into euthymic BALB/cAn (+/+), athymic BALB/cAn (nu/nu) or heterozygous (nu/+) mice.
[b]Tumor size was measured weekly in the mean of two perpendicular diameters with a slide caliper for 4 weeks.
[c]$p < 0.01$, statistically significant by Fisher's exact method against values of Meth A in each group.
[d]$p < 0.05$, statistically significant by Fisher's exact method against values of Meth A in each group.

Remarks

1. The cell wall preparation WPG derived from a strain of *B. infantis* exhibited antitumor effect in a syngeneic transplantable tumor–host system. Histology of the injected site showed tumor cell degeneration during very early time (days 1–3) after injection of the tumor cell–WPG mixture.
2. The early effector mechanism seems not to involve T-cell-dependent mechanism because similar suppression occurred in nu/nu mice, though not to an equivalent efficiency to that in euthymic mice. Preponderance of infiltration of polymorphonuclear leukocyte (PMN) was the main feature of the lesion caused by WPG in mice, suggesting that early inflammatory cell infiltration was the major effector mechanism of this suppression phenomenon.
3. Reimplantation experiments (Expt. III) suggested that the antitumor immunity was acquired in surviving euthymic mice but not in athymic mice, suggesting that the immunological process is also involved in the suppression mechanisms.

Ehrlich Ascites Tumor (13)

Effect of Feeding Yogurt, Milk, and Lactic Acid on Tumor Proliferation

	Tumor Cell ($\times 10^6$/mouse)			DNA (μg/ml) Content		
Material Fed[a]	Control	Test	% Inhibition	Control	Test	% Inhibition
Yogurt	26.3 ± 5.6	20.0 ± 5.3[b]	24	325 ± 7	250 ± 5[b]	23.1
Milk	26.1 ± 3.7	30.6 ± 1.8	0	350 ± 30	370 ± 10	0
Lactic acid	26.5 ± 2.8	28.1 ± 3.5	0	370 ± 8	380 ± 17	0

[a]Yogurt (*L. bulgaricus* and *S. thermophilus*), grade A pasteurized milk (2% skim milk powder) or 1.5% (w/v) lactic acid was fed *ad libitum* to test animals for 7 consecutive days following tumor implantation (injection of 1×10^6 Ehrlich ascites tumor cells, line E-182, generation F-72, obtained from the National Cancer Institute, Bethesda, MD, into peritoneal cavity of each animal). Each test series had 6 male (25–30 g) Swiss mice per group.

[b]Significantly different ($p < 0.05$) from control groups as determined by Student's *t* test.

Remarks

1. Mice fed yogurt for 7 days exhibited 23–24% inhibition of tumor cell proliferation (measured by both cell count and total DNA content).
2. The result indicates antitumor activity of yogurt was due to compound(s) other than lactic acid.

Effect of Prolonged Feeding of Yogurt Diet on Survival of Mice Implanted with Ehrlich Ascites Tumor

		Survival Rate (%)	
	Time (week)	Control	Test
Total of 108 mice were implanted with tumor on day 0. From day 1, animals were given either water (control) or yogurt (test) until death.	2	95	90
	3	50	55
	4	20	23
	5	10	8
	6	3	1
	7	0	0

Remarks Few mice survived more than 4 weeks and percent survival rate was approximately the same for both groups of mice. Yogurt effectively inhibited initial tumor growth but did not appear to retard long-term growth.

Effect of Feeding Yogurt Supernatant Fluid, Solids, and Concentrated Solids on Tumor Cell Proliferation

	Tumor Cell (× 10⁶/mouse)[b]			DNA (μg/ml) Content[c]		
Fraction[a]	Control	Test	% Inhibition	Control	Test	% Inhibition
Supernatant fluid	40.0 ± 2.5	43.7 ± 8.8	0	425 ± 20	430 ± 21	0
Solids	40.0 ± 2.5	26.2 ± 4.7	34.4	425 ± 20	297 ± 23	30.2
3× concentrated solids	28.6 ± 4.8	18.4 ± 2.1	35.6	300 ± 29	195 ± 12	35.0

[a]Yogurt was fractionated by centrifugation. The supernatant fluid and solids were fed separately. Each test series had 6 animals/group. Fed *ad libitum* to animals for 7 days following tumor implantation in peritoneal cavity.

[b]Total number of tumor cells in peritoneal fluid was measured on day 8. The peritoneal fluid was collected with one drop of 20% potassium oxalate added to prevent cell aggregation. Volume of ascite cell suspension was recorded and number of cells counted.

[c]DNA content was measured spectrophotometrically. The cell suspension from each mouse was centrifuged at 610*g* for 10 min at 10°C. The pellet resuspended in 15 ml saline. After homogenization in a Potter Elvehjem tissue grinder, a 1-ml sample was centrifuged at 5900*g* for 10 min at 10°C. The sediment was then washed 3 times with 2.5 ml cold (4°C) 10% trichloroacetic acid (TCA). The 2.5-ml TCA suspension was heated for 15 min at 90°C, centrifuged at 17,600*g* for 10 min and the supernatant fluid mixed with 2.5 ml TCA. Two-tenths milliliters of the supernatant (DNA extract) was mixed with 0.8 ml 2.5% TCA and 2.0 ml dipheylamine reagent (Coleman & Bell, Norwood, OH). The mixture was heated in boiling water bath for 10 min, and the intensity of the blue color determined at 600 nm. A standard curve was prepared with purified calf thymus DNA (Sigma Chemical Co., St. Louis, MO).

Remarks

1. The supernatant fluid fraction of yogurt had no inhibitory effect on tumor growth, while the solid fraction inhibited the tumor by 30–34%.
2. The three times concentrate of the solids showed no significant increase in inhibitory effect, indicating that a maximum level of inhibition may have been reached.

Sarcoma 180 and MCA K-1 Tumor (6)

Effect of Intravenous Administration of LC 9018 on Growth of Sarcoma 180 Inoculated Subcutaneously

Dose[a] (mg/kg × 5)	Tumor Weight (g, mean ± SD)	Inhibition Rate (%)
10	0.64 ± 0.46	75.8
2	0.83 ± 0.50	68.6
0.2	1.52 ± 0.95	42.4
0	2.64 ± 1.57	—

[a]Male ICR mice (20–25 g), purchased from Shizuoka Agricultural Co-operative for Experimental Animal, Hamamatsu, Japan, were inoculated sc with Sarcoma 180 (1 × 10⁶ cells/mouse) on day 0. LC 9018

(heat-killed *L. casei* strain Shirota) was given iv on days +1, +2, +3, +4, and +5. Tumor weights were measured on the 21st day after tumor inoculation.

Remarks Heat-killed *L. casei* strain Shirota (LC 9018) administered iv to ICR mice suppressed the growth of Sarcoma 180 implanted sc.

Antitumor Effect of LC 9018 on MCA K-1 in BALB/c Mice

Dose[a] (mg/kg × 5)	Route	Mean Tumor Wt. (g ± SD)	Inhibition Rate (%)
10	iv	0.99 ± 0.50	57.0
2	iv	0.91 ± 0.62	60.4
0.2	iv	1.72 ± 0.89	25.2
0	iv	2.30 ± 0.88	—
10	ip	0.91 ± 0.68	61.3
2	ip	1.80 ± 0.68	23.4
0.2	ip	1.20 ± 0.96	48.9
0	ip	2.35 ± 0.99	—

[a]Male inbred BALB/c mice (20–25 g) purchased from SACEA, Hamamatsu, Japan, were inoculated sc with MCA K-1 (1×10^6 cells/mouse) on day 0. LC 9018 was given iv or ip on days 1, 2, 3, 4, and 5. Tumor weight was measured on 28th day after tumor inoculation.

Remarks The average tumor weight of mice given heat-killed *L. casei* strain Shirota (LC 9018) ip or iv was significantly lower than that of control mice (similar results were observed *in vitro*).

Dimethylhydrazine- (DMH)-Induced Intestinal Tumor (4)

Effect of *L. rhamnosus* Strain GG on Small Intestinal and Colon Tumor Formation When Fed 2 Weeks before, during, and after Subcutaneous Injection of DMH in Rats Maintained on a Corn Oil Diet

Treatment[a]	*n*	No. of Rats with Tumors (%)	No. of Tumors	Tumors Bearing Animal	Tumor Stage[b]
High-Oil (20%) Diet					
			Small intestine		
DMH	20	9	14	1.55 ± 0.17	6A; 8B
DMH+*L.* GG	21	4[c]	4	1.0 ± 0[d]	2A; 2B
Control	9	0	0	0	
			Colon		
DMH	20	20	74	3.70 ± 0.49	71A; 3B
DMH+*L.* GG	21	15[d]	26	1.73 ± 0.21[e]	26A
Control	9	0	0	0	
Low-Oil (5%) Diet					
			Small intestine		
DMH	21	4	4	1.0 ± 0	4B
DMH+*L.* GG	21	4	4	1.0 ± 0	4A

(continued)

Effect of *L. rhamnosus* Strain GG on Small Intestinal and Colon Tumor Formation When Fed 2 Weeks before, during, and after Subcutaneous Injection of DMH in Rats Maintained on a Corn Oil Diet (*cont.*)

Treatment[a]	*n*	No. of Rats with Tumors (%)	No. of Tumors	Tumors Bearing Animal	Tumor Stage[b]
Control	9	0	0	0	
			Colon		
DMH	20	20	37	1.85 ± 0.25	36A; 1B
DMH+*L.* GG	21	13[d]	19	1.46 ± 0.18	17A; 2B
Control	9	0	0	0	

[a]Male Fisher 344 rats (from Charles River Breeding Lab, Wilmington, MA) were injected subcutaneously every week with 20 mg/kg body wt. of DMH (Sigma Chemical Co., St. Louis, MO) dissolved in 0.9% saline or saline alone (control). Lyophilized lactobacilli (2×10^{11}/100 g diet) were fed to the rats for 16 weeks.
[b]A, tumor is limited to mucosa and submucosa; B, tumor extends into mucosa wall.
[c]$p = 0.073$, using Fisher's Exact Test.
[d]$p < 0.02$, using Fisher's Exact Test.
[e]$p < 0.001$, using Fisher's Exact Test.

Remarks These studies showed that *L. casei* subsp. *rhamnosus* strain GG could interfere with the initiation or early promotional stages of DMH-induced intestinal tumorigenesis; this effect was most pronounced for animals fed a high-fat diet.

Mouse Bladder Tumor-2 (1)

Anti-tumor Activity of *L. casei* Strain Shirota on Mouse Bladder Tumor (MBT-2)[a]

Total Dose (mg/kg)	Route of Administration	Viable/ Killed	No. of Animal	Tumor Wt (g ± SD)	Inhibition (%)	*p*
0 (control)			10	5.09 ± 0.94		
1000 (100 ml × 10 days)	po	V	10	3.20 ± 1.21	37.2	0.01
500 (50 ml × 10 days)	po	V	10	3.61 ± 1.58	29.1	0.05
250 (25 ml × 10 days)	po	V	10	3.42 ± 0.96	32.9	0.01
100 (10 ml × 10 days)	iv	K	9	1.75 ± 0.51	80.4	0.001

[a]Viable or heat-killed *L. casei* strain Shirota (ca. 1×10^8 ml) was given orally (po) or intravenously (iv) to female C3H/He mice (8 weeks old) for 10 days, beginning on seventh day of inoculation of MBT-2 cells (2.78×10^4) into the hind limbs. Tumor which usually occupied the whole extremity of the mouse was resected with the bones and weighed on the twenty-first day.

Remarks Oral administration of viable *L. casei* strain Shirota on seventh day of inoculation for 10 consecutive days resulted in significant inhibitory effect on tumor growth. iv administration of heat-killed bacteria suppressed tumor growth superlatively.

Suppress Secondary Tumor Growth (5)

Treatment[a]	Dose (mg/kg/day)	n	Footpad Thickness (mm, mean ± SEM)[b]		
			Day 7	Day 14	Day 21
Control	0	11	2.93 ± 0.7	5.08 ± 0.19	7.52 ± 0.29
BLP	50	10	2.87 ± 0.07	5.09 ± 0.18	7.21 ± 0.47
BLP	100	10	2.84 ± 0.09	4.26 ± 0.35	5.59 ± 0.68[c]
BLP	200	11	2.62 ± 0.05[d]	3.70 ± 0.30[e]	5.27 ± 0.60[d]

[a]Male BALB/c mice (6–8 weeks old, purchased from Charles River Japan Inc., Kanagawa, Japan) were intradermally inoculated in the abdomen with 5 × 10^5 cells of Colon 26 in saline. Five days later, the tumor mass was surgically removed. Three days after tumor excision, Colon 26 cells (10^5 cells/50 μl saline) were injected into the left-hind footpad. Then BLP (viable *L. casei* strain Shirota) suspension in distilled water was administered orally for 7 consecutive days.
[b]The growth of the tumor implanted was monitored by measuring footpad thickness. The footpad thickness of normal mice was about 2.5 mm.
[c]Significant difference from control: $p < 0.05$, Student's t-test.
[d]Significant difference from control: $p < 0.01$, Student's t-test.
[e]Significant difference from control: $p < 0.001$, Student's t-test.

Remarks

1. Secondary tumor growth in the hind footpad was markedly inhibited in mice that received BLP orally at a dose of 100 or 200 mg/kg/day. Heat-killed *L. casei* strain Shirota (LC 9018) fed orally was notable to suppress the secondary tumor growth.
2. Splenocyte response was also seen to be more responsive to T-cell mitogens and cytokines in the case of the BLP administered mice (data not shown).

Suppress Tumor Recurrence (2)

Characteristics of the Patients with Superficial Bladder Cancer

		All Patients[a]			Subgroups A and B[a]		
		BLP[b]	Placebo[b]	p Value[c]	BLP	Placebo	p Value[c]
Sex	Male	54	51	NS	33	32	NS
	Female	7	13	NS	6	7	NS
Age	15–49	5	6	NS	3	3	NS
	50–59	24	16	NS	15	10	NS
	60–69	17	25	NS	9	14	NS
	70–80	15	17	NS	12	12	NS
Tumor stage	pTa	34	45	$p = 0.128$	20	27	$p = 0.103$
	pT1	21	15	$p = 0.128$	14	8	$p = 0.103$
	pTX	6	4	$p = 0.128$	5	4	$p = 0.103$
Tumor grade	G1	19	31	$p = 0.046$	13	19	NS
	G2	39	30	$p = 0.046$	23	19	NS
	G3	3	3	$p = 0.046$	3	1	NS

(*continued*)

Characteristics of the Patients with Superficial Bladder Cancer (*cont.*)

		All Patients[a]			Subgroups A and B[a]		
		BLP[b]	Placebo[b]	*p* Value[c]	BLP	Placebo	*p* Value[c]
Tumor size, cm	<1	36	46	$p = 0.130$	21	24	NS
	1–3	23	15	$p = 0.130$	16	13	NS
	3–5	2	3	$p = 0.130$	2	2	NS

[a]Study was conducted from September 1990 to November 1992. Patients were registered within 2 weeks of transurethral resection of bladder tumor. Patients were stratified into subgroup A with primary multiple tumor, B with recurrentn single tumors, and C with recurrent multiple tumors. Patients were randomly allocated to receive BLP or placebo.
[b]BLP contained 1×10^{10} viable *L. casei* strain Shirota organisms per gram. Patients were given BLP or a placebo (the vehicle) at a dose of 1 g 3 times per day until tumor recurrence. During the study, no other anticancer therapy was given.
[c]NS, Not significant.

Remarks There was a statistically significant difference in the tumor stage, tumor grade, and tumor size between the BLP and placebo group when the patients were registered for the clinical trial. More patients in the BLP group than in the placebo group had a high risk of tumor recurrence.

Comparison of Cumulative Recurrence-Free Rates for Placebo-Treated and BLP-Treated Patients in Subgroups A and B, after 1 y of Treatment

	n	Recurrence-Free Rate (%)[a]	Corrected Recurrence Free Rate (%)[b]
BLP	39	74.3	79.2
Placebo	39	62.1	54.9
P (Fisher's test)		0.08	0.01

[a]Cytologic examination of urine samples and endoscopy were performed every 3 months after enrollment to detect tumor recurrence, and bladder biopsy was done when necessary.
[b]Data were adjusted as determined by Cox multivariate analysis because of the variations between BLP and placebo groups.

Remarks Cox multivariate analysis showed that the outcome with BLP was significantly better than with placebo. Thus, oral administration of BLP was safe and effective for preventing recurrence of superficial bladder cancer.

Suppress Tumor Metastasis

Lewis Lung Carcinoma

Effect of Postinoculation of *L. casei* after Tumor Inoculation (8)

Group[a]	Adjuvant[b]	Dose	Route of Administration	No. of Pulmonary[c] (Metastases median)	Weight of[c] Lung (mg ± SD)
1	—	Saline × 5	ipl/il	62, 59, 45, 44, 42, 37, 33 (44)	326 ± 52
2	—	Saline × 5	iv + ipl	45, 42, 39, 36, 34, 32, 29 (36)	253 ± 11

(continued)

Effect of Postinoculation of *L. casei* after Tumor Inoculation (8) (*cont.*)

Group[a]	Adjuvant[b]	Dose	Route of Administration	No. of Pulmonary[c] (Metastases median)	Weight of Lung[c] (mg ± SD)
3	LC 9018	250 μg × 5	ipl	24, 21, 20, 19, 16, 15, 15 (19)[d]	240 ± 36
4	LC 9018	250 μg × 5	il	20, 19, 18, 16, 14, 12, 10 (16)[f]	217 ± 30[e]
5	LC 9018	250 μg × 5	il + ipl	14, 12, 9, 8, 7, 5, 4 (8)[f]	202 ± 21[d]
6	LC 9018	250 μg × 5	iv	29, 26, 15, 7, 6, 6, 4 (7)[d]	228 ± 16
7	LC 9018	250 μg × 5	iv + ipl	5, 3, 2, 1, 0, 0, 0, (1)[f]	191 ± 26[d]

[a]Lewis lung carcinoma (3LL) cells (5×10^5/mouse) were inoculated subcutaneously into the left groin of C57BL/6 mice (7 mice/group, 7–10 weeks old, purchased from Shizouka Agricultural Cooperative Association for Laboratory Animals, Hamamatsu, Japan).
[b]Heat-killed *L. casei* strain Shirota (LC 9018) was injected on days 7, 10, 13, 16, and 19.
[c]Number of pulmonary metastases and the weight of the lungs were determined on day 28 after tumor inoculation. The number of metastatic foci was determined by counting the surface colonies.
[d]Statistical significance of difference from control group (group 1 or 2): $p < 0.05$.
[e]Statistical significance of difference from control group (group 1 or 2): $p < 0.01$.
[f]Statistical significance of difference from control group (group 1 or 2): $p < 0.001$.

Remarks Intrapleural, intralesional, iv, and combinations of these three modes of LC 9018 administration were effective in inhibiting pulmonary metastasis after subcutaneous inoculation of Lewis lung carcinoma into mice.

Effect of Priming with *L. casei* Strain Shirota before Tumor Inoculation (11)

Treatment[a]	No. of Pulmonary Metastases (median)	Weight of Primary Tumors (g, mean ± SD)	Inhibition Rate (%)
Control	34, 18, 13, 13, 11 (13)	7.61 ± 1.26	—
LC 9018, intralesional	13, 5, 4, 3, 2 (4)	3.91 ± 1.53[b]	48.6
Priming + LC 9018, intralesional	5, 2, 0, 0, 0 (0)	1.00 ± 0.53[c]	86.9

[a]Heat-killed *L. casei* strain Shirota (LC 9018, 100 μg/mouse) was injected subcutaneously into the right groin of C56BL/6 mice (7–10 weeks old, purchased from Japan SLC Inc. Hamamatsu), 3 weeks and 1 week before tumor inoculation (priming). After Lewis lung carcinoma cells (3LL, 5×10^5/mouse) were inoculated subcutaneously in the left groin of the mice on day 0, LC 9018 (250 μg/mouse) was injected il on days 3, 6, 9, 12, and 15, and the weight of the primary tumors and the number of lung metastases were determined on day 21 after tumor inoculation.
[b]Statistical significance of difference from control: $p < 0.01$.
[c]Statistical significance of difference from control: $p < 0.001$, from LC 9018: $p < 0.01$.

Remarks Intralesional injection of LC 9018 into 3LL-bearing mice inhibited both the growth of the primary tumors and the formation of lung metastases, and this effect was significantly augmented by subcutaneous injection of LC 9018 before the tumor inoculation.

B16 Melanoma (9)

Inhibition of Lung Metastases of Melanoma Inoculated Intravenously

Treatment[a]	Dose	Route of Administration	No. of Lung Metastases[b] (mean ± SD)
Control	Saline × 4	iv	170 ± 36
LC 9018	10 mg/kg × 4	iv	28 ± 16[b]
C. parvum	10 mg/kg × 4	iv	54 ± 10[c]

[a]Cells of highly metastatic variant of the B16-BL6 melanoma cell line (1×10^5 cells/mouse) were inoculated intravenously into C57 BL/mice (7 mice/group, male, 7–10 weeks old, purchased from Shizuoka Agricultural Cooperative Association for Laboratory Animals, Hamamatsu, Japan) on day 0. Heat-killed *L. casei* strain Shirota (LC 9018) or *Corynebacterium parvum* (from Institute Merieux, Lyon, France) was injected iv on days 7, 10, 13, and 16.

[b]Counted on day 21 after cell inoculation. The lung was removed, rinsed in PBS containing heparin (1 unit/ml), and fixed overnight in Bouin's solution. The number of metastatic foci was determined by counting the number of surface colonies.

[c]Statistical significance of difference from control: $p < 0.05$.

[d]Statistical significance of difference from control: $p < 0.01$.

Remarks Injection of heat-killed *L. casei* strain Shirota (LC 9018) iv protected the mice against pulmonary metastasis after iv inoculation of B16-BL6.

Inhibition of Axillary Lymph Node and Lung Metastases and Cytotoxic Activities Against Melanoma

Group[a]	Treatment with LC 9018[b]	Weight of Lymph Node (mg ± SD)[c]	No. of Lumg Metastases (median)[c]
1	—	1770 ± 280	52, 48, 43, 35, 33, 28, 19 (35)
2	il	537 ± 149[d]	42, 40, 38, 30, 29, 17, 16 (30)
3	iv	550 ± 216[d]	35, 23, 21, 13, 11, 10, 5 (13)[e]
4	il and iv	157 ± 14[f]	7, 6, 3, 2, 0, 0, 0 (2)[f]

[a]Melanoma B16-BL 6 cells (2.5×10^5) were inoculated into the left front footpads of C57BL/6 mice (7 mice/group) on day 0.

[b]Heat-killed *L. casei* strain Shirota cells (LC 9018, 10 mg/kg mouse) were injected intralesionally on days 7, 10, 13, and 16 (groups 2 and 4), and the primary tumor was amputated on day 18. After amputation, LC 9018 (10 mg/kg mouse) was injected intravenously on days 20, 23, 26, and 29 (groups 3 and 4).

[c]Axillary lymph nodes were weighed and the pulmonary metastases counted on day 40 after tumor cell inoculation. The weight of the lumph nodes of non-tumor-bearing mice (7 mice/group) on day 40 was 87 ± 4 (mg, mean ± SD).

[d]Statistical significance of difference from control (group 1): $p < 0.05$.

[e]Statistical significance of difference from control (group 1): $p < 0.01$.

[f]Statistical significance of difference from control (group 1): $p < 0.001$.

Remarks Injection of heat-killed *L. casei* strain Shirota (LC 9018) il before surgical excision of the primary tumor inhibited axillary lymph node metastasis. Intravenous injection of LC 9018 after surgical excision of the primary tumor inhibited both axillary lymph node and lung metastases. The combination of il and iv injection of LC 9018 markedly inhibited both lymph node and lung metastases.

REFERENCES

1. Asano, M., Karasawa, E., and Takayama, T. Antitumor activity of *Lactobacillus casei* (LC 9018) against experimental mouse bladder tumor (MBT-2). *J. Urol.* **136:** 719–721, 1986.
2. Aso, Y., Akaza, H., Kotake, T., Tsukamoto, T., Imai, K., and Naito, S. Preventive effect of a *Lactobacillus casei* preparation on the recurrence of superficial bladder cancer in a double-blind trial. *Eur. Urol.* **27:** 104–109, 1995.
3. Goldin, B. R., and Gorbach, S. L. The effect of milk and Lactobacillus feeding on human intestinal bacteria enzyme activity. *Am. J. Clin. Nutr.* **39:** 756–761, 1984.
4. Goldin, B. R., Gualtieri, L. J., and Moore, R. P. The effect of *Lactobacillus* GG on the initiation and promotion of DMH-induced intestinal tumors in the rat. *Nutr. Cancer* **25:** 197–204, 1996.
5. Kato, I., Endo, K., and Yokokura, T. Effects of oral administration of *Lactobacillus casei* on antitumor responses induced by tumor resection in mice. *Int. J. Immunopharmacol.* **16**(1): 29–36, 1994.
6. Kato, I.. Kobayashi, S., Yokokura, T., and Mutai, M. Antitumor activity of *Lactobacillus casei* in mice. *Gann* **72:** 517–523, 1981.
7. Masuno, T., Kishimoto, S., Ogura, T., Honma, T., Niitani, H., Fukuoka, M., and Ogawa, N. A comparative trial of LC9018 plus doxorubicin and doxorubicin alone for the treatment of malignant pleural effusion secondary to lung cancer. *Cancer (Philadelphia)* **68:** 1495–1500, 1991.
8. Matsuzaki, T., and Yokokura, T. Inhibition of tumor metastasis of Lewis lung carcinoma in C57BL/6 mice by intrapleural administration of *Lactobacillus casei. Cancer Immunol. Immunother.* **25:** 100–104, 1987.
9. Matsuzaki, T., Yokokura, T., and Azuma, I. Antimetastatic effect of *Lactobacillus casei* YIT 9018 (LC 9018) on a highly metastatic variant of B16 melanoma in C57BL/6J mice. *Cancer Immunol. Immunother.* **24:** 99–105, 1987.
10. Matsuzaki, T., Yokokura, T., and Mutai, M. Antitumor effect of intrapleural administration of *Lactobacillus casei* in mice. *Cancer Immunol. Immunother.* **26:** 209–214, 1988.
11. Matsuzaki, T., Shimizu, Y., and Yokokura, T. Augmentation of antimetastatic effect on Lewis lung carcinoma (3LL) in C57BL/6 mice by priming with *Lactobacillus casei. Med. Microbiol. Immunol.* **170:** 161–168, 1990.
12. Okawa, T., Niibe H., Arai, T., Sekiba, K., Noda, K., Takeuchi, S., Hashimoto, S., and Ogawa, N. Effect of LC9018 combined with radiation therapy on carcinoma of the uterine cervix. *Cancer (Philadelphia)* **72:** 1949–1954, 1993.
13. Reddy, G. V., Friend, B. A. Shahani, K. M., and Farmer, R. E. Antitumor activity of yogurt components. *J. Food Prot.* **46**(1): 8–11, 1983.
14. Tsuyuki, S., Yamazaki, S., Akashiba, H., Kamimura, H., Sekine, K., Toida, T., Saito, M., Kawashima, T., and Ueda, K. Tumor-suppressive effect of a cell wall preparation, WPG, from *Bifidobacterium infantis* in germfree and flora-bearing mice. *Bifidobact. Microflora* **10**(1): 43–52, 1991.

4.1.8 Prevention of Diabetes (1–3)

Model	Onset (%)[a]	Plasma Glucose[b] (mg/dl)	Plasma Insulin[b] (μU/ml)
KK-A^{y}[c] (3)			
Control	—	566 ± 70	106 ± 25
LC 9018	—	260 ± 105	24 ± 4
NOD[d] (2)			
Control	83	463 ± 254	ND
LC 9018	25	116 ± 4	ND
Alloxan[e] (1)			
Control	100	532 ± 81	ND
LC 9018	40	366 ± 84	ND

[a]Diabetes was diagnosed based on the presence of glycosuria, using glycosuria test paper. Plasma glucose and insulin levels were measured by Glucose B Test Wako (Wako Junyaku, Osaka, Japan) and Elujia-Insulin Kit (Kokusaishinyaku Co., Kobe, Japan), respectively.
[b]Levels of plasma glucose and insulin were measured at 12 (KK-A^{y}) or 30 (NOD) weeks of age or on day 15 after alloxan treatment.
[c]Inbred specific pathogen-free (SPF) male KK-A^{y} mice, a non-insulin-dependent diabetes mellitus (NIDDM) model, at 4 weeks of age were given *L. casei* strain Shirota (2 mg/mouse) orally 5 times a week for 8 weeks. The animals were purchased from CLEA Inc., Tokyo, Japan.
[d]Inbred SPF female nonobese diabetic (NOD) mice, an insulin-dependent diabetes mellitus model, at the age of 4 weeks were given a diet of standard laboratory chow containing 0.05% heat-killed *L. casei* strain Shirota (LC 9018) for 30 weeks. The animals were purchased from CLEA Inc., Tokyo, Japan.
[e]Alloxan at a dose of 70 mg/kg was injected intravenously into inbred SPF 7-week-old male BALB/c mice on day 0, and the mice were given a diet containing 0.05% heat-killed *L. casei* strain Shirota for 15 days. The animals were purchased from SLC, Hamamatsu, Japan.

Remarks Oral administration of *L. casei* strain Shirota (LC 9018) to KK-A^{y} mice, a NIDDM model, decreased both plasma glucose and insulin levels significantly (3). The incidence of diabetes in the NOD mice, an insulin-dependent diabetes mellitus model (IDDM), at the age of 40 weeks was lower in those fed a diet containing 0.05% LC 9018 (3/12) than in the controls given standard laboratory chow (10/12) (2). Feeding mice the LC-containing diet significantly decreased the incidence of diabetes induced by intravenous injection of alloxan, another IDDM model (1).

REFERENCES

1. Matsuzaki, T., Nagata, Y., Kado, S., Uchida, K., Hashimoto, S., and Yokokura, T. Effect of oral administration of *Lactobacillus casei* on alloxan-induced diabetes in mice. *APMIS* **105:** 637–642, 1997.
2. Matsuzaki, T., Nagata, Y., Kado, S., Uchida, K., Kato, I., Hashimoto, S., and Yokokura, T. Prevention of onset in an insulin-dependent diabetes mellitus model, NOD mice, by oral feeding of *Lactobacillus casei*. *APMIS* **105:** 643–649, 1997.

3. Matsuzaki , T., Yamazaki, R., Hashimoto, S., and Yokokura, T. Antidiabetic effects of an oral administration of *Lactobacillus casei* in a non-insulin-dependent diabetes mellitus (NIDDM) model using KK-A^y mice. *Endocr. J.* **44:** 357–365, 1997.

4.1.9 Antihypertensive Effect

Effect of* L. casei *Cell Lysate (LEx) on Systolic Blood Pressure of SHR Rats (1, 2)

	Systolic Blood Pressure (mm Hg, Mean ± S.E.)[b]			
Doses of LEx[a]	Before	6 h	12 h	24 h
0 (control)	180 ± 1	181± 1	180 ± 1	179 ± 1
1 mg/kg	179 ± 1	180 ± 2	176 ± 2	179 ± 1
10 mg/kg	182 ± 1	168 ± 1[c]	167 ± 3[d]	172 ± 2[d]

[a]LEx or distilled water was administered orally to groups of 4–5 animals. *L. casei* strain Shirota cells were harvested, resuspended in distilled water, autolyzed for 2 h at pH 7.0, 55°C, heated at 100°C for 10 min, and then centrifuged. The supernatant was lyophilized and termed LEx.
[b]Male spontaneously hypertensive rats (SHR) purchased from Charles River Japan Inc., were kept at 38°C for a few minutes and systolic blood pressure (SBP) was measured by the tail-cuff method with a programmable sphygmomanometer (type PS-100, Riken Kaihatsu Co., Ltd, Japan).
[c]Significant difference from controls: $p < 0.05$.
[d]Significant difference from controls: $p < 0.01$

Remarks Oral doses of 10 mg/kg of LEx produced a significant SBP decrease in SHR, but LEx had no effect in normotensive rats.

Effect of Cell Wall Polysaccharide–Glycopeptide Complexes of* L. casei *(SG-1) on Systolic Blood Pressure of SHR and RHR Rats (3)

	Systolic Blood Pressure (mm Hg, Mean ± S.E.)[b]			
Doses of SG-1[a]	Before	6 h	12 h	24 h
SHR				
0 (control)	182 ± 2	178 ± 2	179 ± 1	179 ± 2
1 mg/kg	188 ± 1	170 ± 1[e]	170 ± 3[d]	179 ± 2
10 mg/kg	181 ± 1	167 ± 2[e]	166 ± 2[e]	172 ± 2[c]
RHR				
0 (control)	200 ± 7	196 ± 7	193 ± 7	199 ± 8
1 mg/kg	201 ± 6	186 ± 7[c]	183 ± 5[c]	195 ± 7
10 mg/kg	196 ± 11	182 ± 10[c]	178 ± 8	184 ± 9

[a]SG-1 or distilled water was administered orally to groups of 4–8 animals. To prepare for extracts of autologus cell lysates of *L. casei* (LEx) and to isolate SG-1 from LEx, *L. casei* strain Shirota, cells were harvested by centrifugation and suspended in distilled water. Then, cells were autolyzed for 2 h at pH 7.0, 55°C, heated at 100°C for 10 min, and then centrifuged. The lyophilized preparation of the supernatant (LEx, 270 g) was dissolved in 5000 ml of distilled water. After adjusting the pH to 8.0 with 25%

(continued)

ammonium hydroxide, the solution was centrifuged at 14,000*g* for 10 min. The supernatant was filtered through 0.45 μm membrane. Perchloric acid was added to the filtrate to a final concentration of 5% and the precipitate removed by centrifugation at 14,000*g* for 10 min. The precipitate was dissolved in 10 mM ammonium formate buffer at pH 8.5, and perchloric acid was added to a final concentration of 5%, followed by centrifuging once more. These supernatants were mixed and dialyzed against 5 mM ammonium formate buffer (pH 8.0) for 1 day at 4°C. The nondialyzable fraction was filtered through a membrane filter of 0.2 μm. After addition of 500 ml of Q-Sepharose FF to the filtrate to remove proteins and nucleic acids, the nonadsorbed fraction was concentrated to 500 ml by a rotary evaporator. This solution was dialyzed against 100 mM acetate buffer (pH 5.4) supplemented with 0.1 mM zinc chloride. Four thousand units of nuclease P1 was added to the solution and incubated for 5 h at 50°C to digest the nucleic acids. This fraction was digested with 117,000 units of trypsin for 6 h at 37°C. After adjustment of the ammonium formate concentration of this fraction to 1 M (pH 8), it was passed through a column of hydrophobic resin, Phenyl Sepharose CL-4B to remove digestive peptides. The solution passed through the column and was dialyzed against distilled water (MWCO 50,000) for 1 day at 4°C. Finally, the nondialyzable solution was passed through a column of activated charcoal and lyophilized. The fraction obtained was designated SG-1 (3.3 g).

[b]Male spontaneously hypertensive rats (SHR) and renal hypertensive rats (RHR) were kept at 38°C for a few minutes and systolic blood pressure (SBP) was measured by the tail-cuff method with a programmable sphygmomanometer (type PS-100, Riken Kaihatsu Co. Ltd. Japan). The SHR and Wistar-Kyoto rats were purchased from Charles River Japan Inc. To prepare for renal hypertensive rats, Male Wistar-Kyoto rats, 5 weeks of age (body weight, 160–190 g) were anesthetized with sodium pentobarbital (50 mg/kg). A dorsal incision was made, and the left renal artery was stenosed by applying a silver clip (internal diameter, 0.22 mm). The right kidney was left intact. After surgery, the rats were maintained on the standard diet and tap water *ad libitum* for 4–5 weeks. SBP was measured once a week, and rats with SBP above 170 mm Hg were selected for the experiments.

[c]Significant difference from controls: $p < 0.05$.

[d]Significant difference from controls: $p < 0.01$.

[e]Significant difference from controls: $p < 0.001$.

Remarks Antihypertensive compounds were purified from an extract of autologus *L. casei* cell lysates. The most effective compounds were polysaccharide–glycopeptide complexes, found in the cell wall. When these compounds were orally administered to spontaneously hypertensive rats and renal hypertensive rats, SBP decreased by 10–20 mm Hg 6–12 h after administration with no change in heart rate.

REFERENCES

1. Furushiro, M., Hashimoto, S., Hamura, M., and Yokokura, T. Mechanism for the antihypertensive effect of a polysaccharide-glycopeptide complex from *Lactobacillus casei* in spontaneously hypertensive rats (SHR). *Biosci. Biotechnol. Biochem.* **57:** 978–981, 1993.
2. Furushiro, M., Sawada, H., Hirai, K., Motoike, M., Sansawa, H., Kobayashi, S., Watanuki, M., and Yokokura, T. Blood pressure-lowering effect of extract from *Lactobacillus casei* in spontaneously hypertensive rats (SHR). *Agric. Biol. Chem.* **54:** 2193–2198, 1990.
3. Sawada, H., Furushiro, M., Hirai, K., Motoike, M., Watanabe, T., and Yokokura, T. Purification and characterization of an antihypertensive compound from *Lactobacillus casei*. *Agric. Biol. Chem.* **54:** 3211–3219, 1990.

4.2. FARM ANIMALS

4.2.1 Improvement of Growth Rate and Feed Utilization

Animal	Probiotic	Route of Administration	Viability	Quantity	Effect	Ref.
Hubbard chicks	*L. acidophilus*	Oral	Viable	10^{12}/l water	Higher weight gain but slightly lower feed conversion index	8
Young cross-bred pigs	*L. acidophilus*	Oral	Viable	750 mg/kg feed	Increase weight gain and feed conversion	6
	S. faecium	Oral	Viable	1250 mg/kg feed	Decrease weight gain	
Growing-finishing pigs	*L. acidophilus*	Oral	Viable	500 mg/kg feed	No effect	
	S. faecium	Oral	Viable	500 mg/kg feed	No effect	
Piglets	*L. bulgaricus* *L. casei* *S. thermophilus*	Oral	Viable	1 g/kg feed	Consume more feed and grew faster, feed conversion not affected	4
Weaning cross-bred pigs	*Lactobacillus* sp.	Oral	Non-viable	0.18–0.36 ml/kg feed	Improve feed conversion, decrease scouring, greater fiber digestibility and N uptake	2
Growing-finishing cross-bred pigs				0.72 ml/kg feed 0.5–2.0g/kg feed	No effect	
Weaning cross-bred pigs	*Lactobacillus* sp.	Oral	Viable	4×10^6 / kg feed	No effect, some depress daily weight gain	3
Holstein bull calves	*L. bulgaricus*	Oral	Non-viable	1×10^8/day	Higher ration intake and average daily gain	7
Holstein calves	*L. acidophilus* *L. lactis*	Oral	Viable	8% body wt. cultured milk/day	No effect	5
Mixed-breed calves	*L. acidophilus*	Oral	Viable	1.21×10^7/kg feed	Regain weight faster after fasting, higher metabolizable energy	1

Effect on Chicken (8)

	Treatment Groups			
Results	Control	*Lact*[a]	Antib[b]	*Lact* + antib[c]
Mean average weight gain (g)[d]	15.7	17.3	17.8	18.3
Feed consumed/bird/day (g)[e]	24.9	27.2	27.3	27.9

(continued)

Results	Treatment Groups			
	Control	*Lact*[a]	Antib[b]	*Lact* + antib[c]
Feed conversion index	1.59	1.57	1.53	1.52

[a]*L. acidophilus* in drinking water.
[b]Antibiotic in feed.
[c]Lactobacilli in drinking water + antibiotic in feed.
[d]Average of weight gain over period of 7 days.
[e]Average of feed consumed over period of 7 days.

Four groups of 100 Hubbard chicks each were assigned to four treatments: (1) control, (2) *L. acidophilus* in drinking water, (3) antibiotic in feed, (4) *Lact* in water and antibiotic in feed. During the first 5 days, the drinking water for *Lact* and *Lact* + antibiotic groups contained *L. acidophilus* at a concentration of 1 g of freeze-dried product (10^{12} organisms) per liter of water. Each individual bird was weighed daily.

Remarks Weight gain of the bird was much higher in the other diet groups compared to the control as with the case of feed consumed by the bird. However, it was noted that in the feed conversion, there is a tendency toward a higher feed conversion index in the control group.

Effect on Pig

Effect of Probiotics and DL-Lactic Acid on Performance of Pigs (6)

Results	Additive			
	Control (none)	*Lactobacillus*	*S. faecium*	Lactic Acid[a]
Young pigs[b]				
Average daily weight gain (kg)	0.145	0.159	0.134	0.145
Feed-to-gain ratio	3.09	2.43	2.82	2.61
Growing-finishing swine[c]				
Average daily weight gain (kg)	0.84	0.83	0.83	—
Feed-to-gain ratio	3.17	3.20	3.20	—

[a]Lactic acid was not used for the growing-finishing trial.
[b]A total of 224 crossbred pigs (14 pigs/treatment, 2 replications), 4 weeks old were allocated to four different treatment groups: (1) control, (2) Probios (*L. acidophilus,* 750 mg/kg, NuLabs), (3) Feed-Mate 68 (*S. faecium,* 1250 mg/kg, Anchor Labs.), (4) lactic acid (220 mg/kg).
[c]A total of 144 crossbred pigs were used (6 pigs/treatment, 4 replications). The treatment is (1) control, (2) Probios (500 mg/kg), (3) Feed-Mate 68 (500 mg/kg).

Remarks Addition of *L. acidophilus* to the feed of young pigs improved average daily gain by 9.7% and feed conversion by 21.4% whereas the addition of *S. faecium* decreased average daily gain. The addition of lactic acid improved feed conversion, suggesting that lactic acid as a metabolite produced during fermentation might be the cause of improvement in performance. The probiotics had no effect on growing-finishing swine.

Effect of a LFP on Growth, Feed Intake of Weaned Piglets (4)

	Treatment[a]		
Results	Control	LFP	p
Feed intake (kg)	15.78	17.64	0.04
Live body weight (kg)	9.06	10.00	0.04
Feed to gain ratio	1.74	1.75	0.91

Reprinted from Canadian Journal of Animal Science.
[a]Lactobacillus-fermentation-product (LFP, product of Cie Rolmex Inc., Varennes, Quebec) was made of *L. bulgaricus, L. casei* and *S. thermophilus.* A total of 48 piglets weaned between 4–5 weeks of age were put through two treatments of two levels of LFP: none and 0.1% (w/w). Feed intake and body weight were recorded for the first 28 days.

Remarks Piglets fed the LFP diet, consumed 11.8% more feed ($p \leq 0.04$) and grew 10.4% faster ($p \leq 0.04$) than controls, but the feed conversion was not affected. The effect of LFP may be explained by the effect of the microorganisms in LFP on the intestinal flora and by the possible production of metabolites during the fermentation process.

Experiment 1: Weaning Pig (2)

		LFP Level (ml/kg)[b]				
Results[a]	Control	0.18	0.36	0.72	1.44	SE[c]
Weight gain (kg)	10.45	11.01	11.01	10.96	10.64	± 0.24
Feed gain[d] (kg/kg)	2.20	2.02	1.99	2.06	2.17	± 0.035
Scour index[e]	3.9	4.3	4.5	4.55	4.4	± 0.065

[a]Five groups of six cross-bred weaning pigs (4–5 weeks) were used. The trial took 28 days.
[b]Basal diet was supplemented with liquid LFP, which contained nonviable *Lactobacillus* species.
[c]Standard error of the mean.
[d]Calculated from the weight gain and feed intake.
[e]Scour index is an average of degree of scouring recorded daily during first 14 days of trial where 1 = severe scouring and 5 = none.

Remarks Pigs fed a diet with 0.36 ml/kg LFP required nearly 10% less feed per unit of weight gain than the control pigs. Also the incidence of scouring decreased ($p < 0.05$) in pigs fed with different levels of LFP. Overall improvement occurred up through the addition of 0.36 ml/kg LFP with no additional benefit from greater amounts.

Experiment 2: Growing Pig (2)

Item[a]	Basal Diet	LFP Supplemented Diet[b]	SE[c]
Dry matter			
Apparent absorption %	79.8	79.5	± 0.43
Crude fiber			
Apparent absorption %	39.9[e]	45.6[e]	± 0.94

(continued)

Experiment 2: Growing Pig (2) (*cont.*)

Item[a]	Basal Diet	LFP Supplemented Diet[b]	SE[c]
NFE[d]			
Apparent absorption %	84.8	84.7	± 0.41
Nitrogen (g)			
Intake	210.9	209.1	± 5.4
Fecal	45.6	47.0	± 2.3
Urinary	64.5	63.8	± 4.0

[a]Mean of 12 observations per diet for 19 days. On the last 5 days, total, fecal, and urinary collections were done, after which the 2 groups had their diets switched around where the group on a basal diet was now fed the supplemented one and vice versa. Period length and collection of feces and urine were the same as pervious experiment.
[b]Basal diet supplemented with 0.72 ml liquid LFP/kg.
[c]Standard error of the mean.
[d]Nitrogen free extract.
[e]Means on same line are significantly different ($p < 0.05$).

Remarks For pigs fed the diet supplemented with LFP containing nonviable *Lactobacillus* sp., the digestibility of fiber was greater ($p < 0.05$) than pigs fed the basal diet. It could be explained that possibly the rate of passage of feed through the digestive tract was decreased by feeding LFP, which allowed more time for digestion of crude fiber. Also, the urinary nitrogen excretion was greater than fecal excretion but both combined were less then the intake, thus resulting in a positive nitrogen balance.

Experiment 3: Growing-finishing pig (2)

Results[a]	Treatment[b] 1	2	3	4	5	SE[c]
Ave. weight gain (kg/day)	0.76	0.78	0.79	0.79	0.81	± 0.02
Feed gain (kg/kg)	3.00	2.95	2.96	3.01	3.00	± 0.04

[a]Eight individually fed pigs/treatment, 97 days trial.
[b]Treatments: 1, control: no LFP added; 2, 0.72 ml liquid LFP/kg diet, the LFP containing nonviable *Lactobacillus* sp.; 3, 0.5 g dry LFP/kg diet (dry LFP was obtained by mixing liquid LFP with a wheat bran carrier and dried at about 40°C); 4, 1.0 g dry LFP/kg diet; 5, 2.0 g dry LFP/kg diet.
[c]Standard error of mean.

Remarks LFP when included in the diet of growing-finishing pigs did not significantly affect ($p > 0.005$) rate of gain or feed efficiency.

Effect of Probiotics Supplementation on the Performance of Weaned Pigs and Growing-Finishing Swine (3)

	Treatment		
Results	Control	Probiotic[a–c]	SE[d]
Trials 1–3[e]			
Daily gain (g)	322	317	± 6.40
Daily feed (g)	625	616	± 12.71
Feed/gain (kg/kg)	1.94	1.94	± 0.024
Trials 4 + 5[f]			
Daily gain (g)	590	556[g]	± 9.08
Daily feed (kg)	1.88	1.83	± 0.039
Feed/gain (kg/kg)	3.19	3.29	± 0.049
Trials 6–8[h]			
Daily gain (kg)	0.73	0.74	± 0.012
Daily feed (kg)	2.17	2.22	± 0.043
Feed/gain (kg/kg)	2.99	2.99	± 0.041

[a]*Lactobacillus* probiotic (contained 4×10^6 viable cells per gram, Pioneer Hi-Bred International, Portland, OR 97207) at 1 g/kg of diet for trials 1–3.
[b]*Lactobacillus* probiotic at 1 g/kg of diet to 20 kg of body weight then at 500 mg/kg of diet for trials 4 + 5.
[c]*Lactobacillus* probiotic at 500 mg/kg of diet for trials 6–8.
[d]Average standard error of the mean.
[e]Each mean represents 22 pens of 6–8 weaned cross-bred pigs per pen.
[f]Each mean represents 6 pens of 5–6 weaned pigs per pen.
[g]Probiotic lower ($p < 0.05$) than control.
[h]Each mean represents 10 pens of 5–6 growing-finishing pigs/pen.

Remarks

1. Trials 1–3: Supplementation of diet with 1g/kg *Lactobacillus* probiotic did not improve daily gain, feed intake, or feed efficiency.
2. Trails 4–5: Daily gain of pigs fed on diets supplemented with *Lactobacillus* probiotic were lower ($p < 0.05$) than pigs on the control diet, but differences in feed intake or feed efficiency were not significant.
3. Trials 6–8: *Lactobacillus* probiotic supplementation showed nonsignificant trends for improved daily gain and feed intake over the control.

Overall, the addition of probiotic to diets of weaned pigs, grower and finisher did not improve growth performance or feed efficiency. In trials 4 and 5, the daily gain was depressed with *Lactobacillus* probiotic supplementation. No explanation for this depressed gain is available.

Effect on Cattle

Performance of Calves Fed Incremental Nonviable LFP (7)

		LFP (ml/day)				
Results[a]	Control	3	6	12	24	SE[b]
Average daily gain, kg[c]						
0–4 wk	0.22	0.26	0.22	0.29	0.25	0.04
4–11 wk	0.76	0.77	0.76	0.84	0.76	0.05
0–11 wk	0.56	0. 59	0.56	0.64	0.57	0.04
Pelleted ration intake, kg dry matter/day[c]						
0–4 wk	0.17	0.26	0.17	0.24	0.22	0.03
4–11 wk	1.91	2.09	1.93	2.10	1.94	0.12
0–11 wk	1.27	1.43	1.29	1.42	1.31	0.08
Gain/feed, kg/kg dry matter[c]						
0–4 wk	0.40	0.40	0.38	0.47	0.39	0.06
4–11 wk	0.38	0.36	0.37	0.38	0.38	0.01
0–11 wk	0.38	0.36	0.38	0.39	0.38	0.01

[a]Forty Holstein bull calves were assigned randomly to five different quantities of liquid LFP containing 10^8 nonviable *L. balgaricus* per ml liquid (TransAgra Corp., Memphis, TN); 0, 3, 6, 12, or 24 ml per day. Feed consumption was measured daily while the calves were weighed weekly.
[b]Standard error for $n = 8$.
[c]Mean adjusted by covariance to average initial weight of 42.6 kg.

Remarks Feeding LFP did not affect significantly intake of pelleted ration and average daily gain (ADG). It was noted that the pelleted ration intake was higher for calves receiving either 3 or 12 ml/day LFP while the ADG was higher for calves fed 12 ml/day LFP as compared to the control group.

Performance of Calves Fed Cultured Milk (5)

		Weekly Starter Consumption (kg)				
Treatment[a]	Weekly Gain (kg)	1	2	3	4	5
Control	3.19	0.06	0.92	2.34	4.13	5.17
Cultured milk	2.82	0.05	0.77	1.95	2.72	4.27

[a]Holstein calves in four groups were given treatment of either pasteurized milk (control) or cultured milk, fed at 8% of body weight in two equal feedings daily during the first week and once daily thereafter. Cultured milk was prepared by inoculating pasteurized milk with *L. acidophilus* and *L. lactis* and incubation at 37°C for 24 h, then refrigerating at 10°C until used. Body weights were recorded weekly. Duration of trial was 5 weeks.

Remarks The cultured milk did not affect weight gains significantly. Calves fed cultured milk consumed less starter ($p = 0.0004$).

Daily Dry Matter and Metalizable Energy Intakes of Calves Fed Live *L. acidophillus*

			Refed Period (day)				
Treatment[a]	Prefast	Fast	1	2–3	4–8	9–14	0–14
Dry matter, kg/d							
Control	10.2	9.8	9.7	9.9[b]	9.6	9.1	9.4
HR	10.2	2.0	5.5[b]	7.7	7.7[b]	9.9	8.6
LAC	10.2	2.2	8.6	7.7	9.0	9.7	9.1
SE[c]	0.86	0.72	0.96	0.58	0.54	0.69	0.48
Calculated metabolizable energy, m cal/d							
Control	18.2	17.5[b]	17.4	17.7	17.2[b]	16.2	16.8
HR	18.2	3.6	9.8[b]	13.8[b]	13.8[b]	17.7	15.4
LAC	18.2	3.9	21.5	19.2	22.5[b]	24.2[b]	22.8[b]
SE[c]	1.53	1.28	2.25	1.54	1.41	1.82	1.30

[a]Six groups of 10 mixed-breed calves were used. Five groups were deprived of feed and water for 24 h, limit-fed hay for 24 h, deprived for 48 h and then limit-fed (4.5 kg/head) one of the five diets daily for 2 weeks (realimentation period). Calves in the last group were fed high-roughage diet throughout the experiment. Calves were individually weighed immediately before the start of the deprivation period, at the end of the deprivation period and on days 1, 3, 8, and 14 of the realimentation period. Control = not fasted, HR = high roughage, LAC = *L. acidophilus,* 5.5×10^6 CFU/g × 2.2 g/kg feed.
[b]Values in same column that have a superscript differ significantly from the others ($p < 0.05$).
[c]Standard error of the mean.

Remarks Calves that were fed the LAC diet regained their weight faster than that of the HR diet. Calves in the LAC group also had a higher calculated metabolizable energy compared to the control and the HR group.

REFERENCES

1. Cole, N. A., and Hutcheson, D. P. Influence of realimentation diet on recovery of rumen activity and feed intake in beef steers. *J. Anim. Sci.* **61**(3): 692–701, 1985.
2. Hale, O. M., and Newton, G. I. Effects of a nonviable *Lactobacillus* species fermentation product on performance of pigs. *J. Anim. Sci.* **48**(4): 770–775, 1979.
3. Harper, A. F., Kornegay, E. T. Bryant, K. L., and Thoman, H. R. Efficacy of virginiamycin and a commercially-available *Lactobacillus* probiotic in swine diets. *Anim. Feed Sci. Technol.* **8:** 69–76, 1983.
4. Lessard, M., and Brisson, G. J. Effect of a *Lactobacillus* fermentation product on growth, immune response and fecal enzyme activity in weaned pigs. *Can. J. Anim. Sci.* **67:** 509–516, 1987.
5. Morrill, J. L., Dayton, A. D., and Mickelsen, R. Cultured milk and antibiotics for young calves. *J. Dairy Sci.* **60**(7): 1105–1109, 1977.
6. Pollmann, D. S., Danielson, D. M., and Peo, E. R., Jr. Effects of microbial feed additives on performance of starter and growing-finishing pigs. *J. Anim. Sci.* **51**(3): 577–581, 1980.
7. Schwab, C. C., Moore, J. J. III, Hoyt, P. M., and Prentice, J. L. Performance and fecal flora of calves fed a nonviable *Lactobacillus bulgaricus* fermentation product. *J. Dairy Sci.* **63:** 1412–1423, 1980.

8. Tortuero, F. Influence of the implantation of *Lactobacillus acidophilus* in chicks on the growth, feed conversion, malabsorption of fats syndrome and intestinal flora. *Poultry Sci.* **52:** 197–203, 1973.

4.2.2 Improvement of Disease Resistance

Animal	Probiotics	Route of Administration	Viability	Concentration	Effect	Ref.
Weaned bull calves	*B. pseudolongum*	Oral	Viable	1 g/day	Decrease in occurrence of diarrhea	1
Suckling piglets	*B. pseudolongum*	Oral	Viable	0.5 g/day	Decrease in occurrence of diarrhea, better health	1
Weaned piglets	*L. bulgaricus*	Oral	Viable	500 ml cultured milk/l diet or 36 g dry cultured milk/l diet	Lower percentage of hemolytic *E. coli* in rectal swabs, and delay onset of mortality	2

Prevention of Diarrhea (1)

Prevention of Diarrhea in Calves

	Trials–Scouring Calves				
Treatment[a]	1	2	3	4	5
B. thermophilum					
Nontreated	2	9	12	11	6
Administered	0	4	7	7	3
B. pseudolongum					
Nontreated	16	8	4	—	—
Administered	5	3	0	—	—

[a]A total of 800 early weaned bull calves (100 calves per treatment) were given VBP (viable bifidobacteria preparation). They were fed twice a day with 250 g of commerical powdered milk replacer. The animals were given 1 g of VBP of *B. pseudolongum* ($10^{9.3}$ CFU/g) or *B. thermophilum* ($10^{8.3}$ CFU/g) suspended in milk replacer, once a day for 20 days. Two groups were used as control. Over 30 days from the beginning of the administration, occurrence of diarrhea along with health condition were examined. Calves that scoured for 3 consecutive days or more were regarded as "scouring."

Remarks There is a significant decrease in occurrence of diarrhea in calves fed *Bifidobacteria* compared to control groups (chi-square test: $p < 0.01$, $p < 0.001$)). The incidence of diarrhea was 8.4% (21/250) for *B. thermophilum* treatment group and 5.3% (8/150) for *B. pseudolongum* treatment group as compared to 16.0% (40/250) and 18.7% (28/150) in the control group.

Prevention of Diarrhea in Pigs

	No. of Piglets Scoured[a]	
Ages in Days	Control Groups	*Bifidobacteria* Administrated Groups
In winter	59 piglets	64 piglets
1–7	41 (69.5%)	2 (31%)
8–14	11 (21.2%)	9 (14.3%)
15–21	4 (8.2%)	4 (6.5%)
In summer	79 piglets	110 piglets
1–7	48 (60.8%)	5 (4.5%)
8–14	17 (23.3%)	23 (20.9%)
15–21	8 (11.4%)	6 (5.5%)

[a]A total of 312 piglets were given VBP (123 piglets in winter, 189 piglets in summer). VBP (*B. pseudolongum* preparation) was given orally at a dose at 0.5 g suspended in water once a day for 10 days starting immediately after birth. Incidence of diarrhea was examined for 3 weeks after birth.

Remarks Diarrhea occurred scarcely in the VBP groups during the first week of birth, while more than 60% of piglets in the control groups suffered from diarrhea. In general, diarrhea occurred less frequently in the VBP groups than in the nontreated control group during the suckling period (1–21 days old). Animals treated with VBP were better in health and growth than the control group.

***Neutralize Enterotoxin from* E. coli (2)**

Treatment[a]	n^b	Avg. (%) hemolytic *E. coli*[c]	Avg. Days of Scour	% Survival	Avg./Days of Death for Pigs Dying
Mild challenge (< 20% mortality)[c]					
Control (1)[d]	26	15.7	1.3	100	—
Experimental (1)[e]	27	14.1	0.9	100	—
Moderate challenge (20–60% mortality)[c]					
Control (1)[d]	30	20.1	3.9	43.3	6.5
Experimental (1)[e]	34	20.8	4.6	67.6	7.1
Control (2)[f]	33	34.2	2.2	66.7	8.2
Experimental (2)[g]	33	32.1	1.7	66.7	8.6
Severe challenge (> 60% mortality)[c]					
Control (1)[d]	16	64.0	3.0	12.5	5.8
Experimental (1)[e]	16	47.0	4.0	12.5	7.6

[a]Feeding experiments were of 2 weeks duration. Rectal swabs taken on days 1, 5, 9, 12, and 14. Swabs were streaked out on blood agar and percentage of hemolytic *E. coli* determined after 24 h growth at 37°C. Scouring was assessed visually.

[b]Number of pigs in trial group.
[c]*E. coli* G7, G1253 obtained from Central Veterinary Lab., Weybridge, Surrey, UK; P16, P99, and P115 obtained from Houghton Poultry Research Station, Houghton, Hunts, UK, were given to the piglets on 2 consecutive days after weaning.
[d]Autoclaved liquid *L. bulgaricus* FW 148 obtained from Unilever Research Lab., Welwyn, Herts, UK, 500 ml cultured milk/l diet.
[e]Liquid *L. bulgaricus* broth, 500 ml cultured milk/l diet.
[f]Spray dried skim milk, 36 g/l diet.
[g]Spray dried *L. bulgaricus* broth, 36 g/l diet.

Remarks The feeding of *L. bulgaricus* in the diets help to lower the percentage of average hemolytic *E. coli* and the delay of the onset of mortality compared with the controls.

REFERENCES

1. Kimura, N., Yoshikane, M., Kobayashi, A., and Mitsuoka, T. An application of dried *Bifidobacteria* preparation to scouring animals. *Bifidobact. Microflora* **2**(1): 41–55, 1983.
2. Mitchell, I. De G., and Kenworthy, R. Investigation on a metabolite from *Latobacillus bulgaricus* which neutralizes the effect of enterotoxin from *Escherichia coli* pathogenic for pigs. *J. Appl. Bacteriol.* **41:** 163–174, 1976.

5

MECHANISMS OF PROBIOTICS

5.1 COMPETITIVE EXCLUSION

Pathogen	Subject	Probiotic	Route of Administration	Viability	Ref.
Escherichia coli	Large White pigs	*Lactobacillus fermentum* strain 14	Oral	Viable	1
E. coli Ecb028	Grey Leghorn chicks	*L. acidophilus* group B strain 59	Oral	Viable	5
Salmonella Kedougou NCTC12173	Chicks	Indigenous intestinal flora	Oral	Viable	3
Salmonella typhimurium	White Rock × Cornish chicks	Indigenous intestinal flora	Oral	Viable	4
Uropathogens	Human uroepithelial cells	Urethral *Lactobacillus* isolate	—	Viable	2

5.1.1 Competitive Exclusion of *E. coli*

Experimental Protocol (1)

1. *L. fermentum* strain 14 and *Streptococcus salivarius* strain 32 were inoculated in 4 l MRS broth and 500 ml YGB, respectively, and incubated overnight at

37°C. The cultures were centrifuged at 1500g for 30 min, the deposits resuspended in sterile PBS and adjusted to 4×10^{10} bacteria/ml.

2. For the first 24 h after weaning the Large White pigs were fed by hand every 8 h with 2 ml of the concentrated suspension containing 4×10^{10} lactobacilli/ml and 8×10^{7} streptococci/ml. Hand feeding was done with a 5-ml plastic syringe with 4 cm of silicone rubber tubing attached.
3. Thereafter, the probiotics (ca. 5×10^{8} lactobacilli and 1×10^{5} streptococci/ml) were included in the milk substitute corresponded to the numbers previously found in the stomach of suckling pigs. Suspensions containing 5×10^{9} lactobacilli/ml and 1×10^{6} streptococci/ml were placed in inverted 500-ml blood bottles fitted with a rubber stopper with a short glass outlet tube and a long inlet tube for air intake.
4. The bacterial suspension and milk diet in the ratio of 1:10 was delivered simultaneously into the feeding trough using the automatic feeding mechanism. The bulk suspension was changed every 24 h. Stock suspensions were stored at 4°C for the duration of the experiments.
5. Four days after weaning the pigs were killed and bacterial counts made on the stomach and duodenum contents.

			$\log_{10}$ Viable Count of *E. coli*/g			
Feeding Regimen[a]	Site Examined	No. of Comparisons	Control Pigs	Probiotic Fed Pigs	Change in Value ± SE	*p*
L. fermentum and *S. salivarius* for 4 days	Stomach	21	7.37	6.04	−1.33 ± 0.40	< 0.01
	Duodenum	19	7.02	6.25	−0.77 ± 0.33	< 0.05
L. fermentum for 4 days[b]	Stomach	6	7.75	6.79	−0.96 ± 0.30	< 0.05
	Duodenum	6	7.21	6.21	−1.00 ± 0.49	NS[c]
S. salivarius for 4 days[b]	Stomach	6	7.89	7.73	−0.16 ± 0.04	NS
	Duodenum	7	7.31	7.42	+0.11 ± 0.27	NS

Reproduced with permission of Blackwell Science Ltd.

[a]The *in vivo* antagonistic effect of adhering bacteria for *E. coli* was tested by feeding selected strains to artificially reared pigs, which have a high count of *E. coli.*

[b]Strains from several sources were tested for adhesion *in vitro,* but only those from pigs and chickens attached. Of the isolates tested, *L. fermentum* strain 14 and *S. salivarius* strain 312 were chosen for this study for the following reasons: (1) They were isolated more frequently from the anterior gut of suckling pigs where the count of *E. coli* was lower than from early-weaned pigs, where it was high. (2) They were regularly isolated from the nonsecretory region (pass esophagus) of the stomach. (3) In the *in vitro* adhesion test, they attached better than any of the other isolates.

[c]NS, not significant.

Remarks A statistically significant reduction in the *E. coli* count was produced in both the stomach and duodenum by feeding a combination of *L. fermentum* 14 and

S. salivarius 312 for 4 days. When *L. fermentum* alone was fed, there was also significant reduction in *E. coli* numbers in the stomach. In cases of diarrhea caused by *E. coli,* the treatment as described here was not effective because the count of these organisms in the duodenum of culture-fed pigs was still $>10^6$/g. If, however, the antibacterial effect of strain 14 could be increased, some effect on scouring due to *E. coli* should follow. This might be accomplished by the feeding of large numbers of organisms or by the administration in a concentrated form of the inhibitory factor(s) produced by *L. fermentum* strain 14.

Effect of Subsequent *L. acidophilus* Dosing on Gut Shedding of *L. acidophilus* and *E. coli* in Gnotobiotic Chicks (5)

Treatments[a]	Dose Regimen[a]	Ave % Collection of *L. acidophilus*[b]	Ave % collection of *E. coli*[b]	Total Chicks (start/finish)
E. coli (therapeutic)				
E	1	0	100	113/78
E–L	1–1	34	100	
E–L	1–3	60	98	
E–L	1–4	63	99	
E–L	1–5	76	93	
Group average		46	88	
L. acidophilus (prophylactic)				
L	1	100	0	108/103
L–E	1–1	72	100	
L–E–L	1–1–3	83	68	
L–E–L	1–1–4	96	48	
L–E–L	1–1–5	99	47	
Group average		89	52	
Population average		68	70	

[a]All gnotobiotic Grey Leghorn chicks (obtained from Colorado State University breeder flock) were dosed at 2 days of age with either *L. acidophilus* group B strain 59 or pathogenic *E. coli* EC B028. Additional doses of either bacterium followed the dose regimes as listed in table, which were administered every 2 days after initial dosing. All doses were given orally. L denotes a dose of *L. acidophilus* and E denotes a dose of *E. coli.* The dose regimen indicates the number of doses of each that were given. Each dose consisted of the test organism containing between 10^8 to 10^9 organisms/ml.

[b]Cloacal swabs were collected from gnotobiotic chicks to observe the shedding of bacteria from the alimentary tract. A total of six swabs were obtained from each chick on days 4, 6, 8, 10, 12, and 14.

Remarks When *L. acidophilus* was used therapeutically, it did not reduce pathogenic *E. coli* shedding even with subsequent *L. acidophilus* dosing. Prior inoculation of germ-free chicks with *L. acidophilus* reduced shedding of pathogenic *E. coli* from 100 to 47%. The *L. acidophilus* was competitive with *E. coli,* since it prevented the colonization and therefore decreased the shedding of *E. coli* by 53%.

Frequency of Isolating *L. acidophilus* (L) or *E. coli* (E) from Gut of Gnotobiotic Chicks at Postmortem (5)

Treatment	Trial	Dose Regimen	Total Chicks (start/finish)	L/E Frequency by Gram Stain[a]				
				C	P	D	Ce	R
NC[b]	11		7/6	0/0	0/0	0/0	0/0	0/0
NC	1		4/2	0/0	0/0	0/0	0/0	0/0
Therapeutic								
E	1	1	3/0	—	—	—	—	—
E	11	1	6/4	0/3	0/3	0/3	0/3	0/3
E–L	2	1–1	9/7	5/7	4/7	1/7	5/7	3/7
E–L	6	1–1	8/7	3/6	4/4	1/6	1/7	2/6
E–L	8	1–3	15/9	9/9	8/8	6/5	9/8	9/9
E–L	3	1–3	19/6	6/5	5/6	5/6	3/6	5/6
E–L	7	1–4	11/4	4/3	4/2	4/4	4/4	4/4
E–L	7	1–4	16/15	15/14	15/13	12/13	10/15	13/15
E–L	9	1–5	10/10	10/10	10/7	8/7	10/8	9/8
E–L	5	1–5	16/16	14/14	15/14	14/14	12/15	11/16
Prophylactic								
L	12	1	5/5	5/0	5/0	5/0	5/0	5/0
L	1	1	4/4	4/0	4/0	4/0	4/0	4/0
L–E	10	1–1	9/9	9/9	9/9	7/9	5/9	6/9
L–E	2	1–1	7/7	7/7	7/7	5/7	6/7	4/7
L–E–L	3	1–1–3	17/15	15/15	15/12	14/12	14/10	15/11
L–E–L	10	1–1–3	15/14	13/14	11/13	12/10	13/12	11/14
L–E–L	5	1–1–4	10/9	9/5	9/6	9/7	9/6	9/8
L–E–L	8	1–1–4	12/11	11/8	11/6	11/10	11/8	11/11
L–E–L	6	1–1–5	14/14	14/8	14/14	14/11	14/12	14/10
L–E–L	9	1–1–5	15/15	15/9	15/8	15/10	15/11	15/13

[a]Gnotobiotic chicks were not administered any microbial cultures.
[b]Chicks killed and gut contents from five different segments of the alimentary tract that included the crop (C), proventriculus (P), duodenum (D), cecum (Ce), and rectum (R) were examined. Inoculum from each section were incubated and gram staining done on these cultures to differentiate the presence of *L. acidophilus* and *E. coli*.

Remarks When gnotobiotic chicks were sacrificed, *L. acidophilus* was observed throughout the gastrointestinal tract, indicating that this strain of *L. acidophilus* can withstand passage through the gastric proventriculus and colonize in the presence of bile. Chicks maintained on the prophylactic treatments had fewer *E. coli* than chicks on the therapeutic treatments. *L. acidophilus* was more effective when used prophylactically to prevent *E. coli* from populating the gastrointestinal tract.

5.1.2 Competitive Exclusion of *Salmonella*

Data for 10 Control and 10 Treated Chicks[a] (3)

Log_{10} Salmonellae/g	No. of Positive Chicks at Each Level[b] Control Group[c]	Treated Group[d]
7	4	0
6	2	0
5	1	0
4	0	0
3	2	0
2	0	0
1	0	0
Not found (0)	1	10

[a]Newly hatched chicks of either sex were obtained from hatchery. The chicks were tested to ensure they were initially free from *Salmonella*.
[b]From each subgroup, a minimum of five birds were taken at random, killed, and examined 5 days after challenge to determine both the proportion of positive birds in treated and control groups and the levels of *Salmonella* carriage in infected individuals.
[c]Chicks in this group did not receive live culture treatment before challenge with *Salmonella kedougou* NCTC 12173. Live culture was suspensions or cultures of gut content from suitable adult "donor" birds or defined mixtures of intestinal bacteria.
[d]Chicks in this group were treated with live culture before challenge with *Salmonella* 24 h after being given the protective treatment by allowing them to drink 0.5 ml from the tip of a pipette. The *Salmonella* used was a nalidixic acid-resistant *Salmonella kedougou* (NCTC 12173). The strain was sensitive to antibiotics. The *Salmonella* was used at a challenge level of ca. 10^4 organisms/chick. The strain was noninvasive but readily colonizes the ceca.

Remarks With live culture (taken from adult donor birds) administration, there was rapid establishment of an adult intestinal microflora in young chicks that increased their resistance to colonization by food-poisoning Salmonellae and provided a possible means of reducing *Salmonella* infection in commercial poultry flocks.

Comparison of Protective Activity of Cultures from Ceca and Feces of Adult Donor Birds Against *Salmonella* (4)

Grams of Inoculum/ Treatment Culture[a]	Washed Ceca			Whole Ceca			Feces		
	Percent Infection	Infection Factor[b]	Protection Factor[c]	Percent Infection	Infection Factor	Protection Factor	Percent Infection	Infection Factor	Protection Factor
10^{-3}	5	0.05	>25.0	0	<0.1	>25.0	10	0.25	>25.0
10^{-4}	5	0.05	>25.0	0	<0.1	>25.0	5	0.05	>25.0
10^{-5}	37	0.9	4.9	5	0.05	>25.0	10	0.1	>25.0
10^{-6}	25	0.8	5.5	5	0.05	>25.0	5	0.2	>25.0
10^{-7}	30	1.4	3.1	5	0.05	>25.0	20	0.8	8.3

(*continued*)

Comparison of Protective Activity of Cultures from Ceca and Feces of Adult Donor Birds Against *Salmonella* (4) (*cont.*)

Grams of Inoculum/ Treatment Culture[a]	Washed Ceca			Whole Ceca			Feces		
	Percent Infection	Infection Factor[b]	Protection Factor[c]	Percent Infection	Infection Factor	Protection Factor	Percent Infection	Infection Factor	Protection Factor
10^{-8}	57	1.9	2.3	35	1.6	3.5	35	2.0	3.3
Nontreated (control)	95	4.4		100	5.6		100	6.6	

[a]Each 3-day-old culture in VL medium (37°C) from washed ceca, whole ceca, and feces of adult donor birds was tested for protective activity against 10^5 CFU of *Salmonella typhimurium* by introducing 0.5 ml of a 1:10 dilution into the crop of 10 1-day-old chicks (White Rock × Cornishy, supplied by Curtis Chick Ltd., Port Hope, Ontario). Results are average of 2 trials, each trial with 20 chicks.
[b]Infection factor (the mean $\log_{10}$ value of the number of *Salmonella*/g of cecal content for a group of chicks treated and challenged identically). Infection factor takes into account the degree of infection.
[c]Protection factor (ratio of infection factor for untreated/treated chicks). A protection factor value exceeding 25.0 (>25.0) signifies essentially full protection while a value of <4.0 shows low or no protection.

Remarks Whole ceca and feces contain a larger assortment and number of bacteria than washed ceca. As inoculum size was decreased, there was gradual loss of protective activity against *Salmonella* for all three cultured preparations indicating that not all species of bacteria needed for starting fully protective cultures were present. The results with washed ceca at low dilutions of inoculum showed that bacteria relevant to protection resisted the washing procedure and presumably remained adherent to the cecal wall.

Protection of Washed Ceca Cultures Against *Salmonella* in Chicks That Had Been Given Fecal Cultures for Various Duration (4)

Duration of Passage (h)[a]	Percent Infection	Infection Factor[b]	Protection Factor[b]
0	85	3.9	1.1
0.5	31	0.7	6.1
1	23	0.7	6.1
3	20	0.5	7.7
6	8	0.2	21.3
8	0	0	>25.0
18	0	0	>25.0
24	5	0.2	21.3
48	15	0.2	21.3
72	3	<0.1	>25.0
336	8	0.1	>25.0
Nontreated[c] (ceca washed at 48 h)	85	4.1	1.0
Control	90	4.3	

[a]Fecal culture was passaged for a certain time in newly hatched chicks, than the ceca were excised, washed, homogenized, and cultured for 72 h in VL medium. The cultures were assayed for anti-Salmonella activity in 1-day-old chicks.

[b]See footnotes *b* and *c* of the previous table.
[c]The same as in footnote *a*, but chicks were not given fecal cultures.

Remarks Cultures from washed ceca of chicks sacrificed 30 min or 1 h after administration of fecal culture had the same protective activity (PF 6.1), and that the protective flora started to adhere to the cecal wall almost immediately. After 6–8 h, all species of bacteria necessary for starting protective culture appeared to be attached.

5.1.3 Competitive Exclusion of Uropathogenic Bacteria (2)

Inhibition of Uropathogens Adherence by *Lactobacillus* Whole Cells and Various Cell Wall Fragments

	% Inhibition (mean ± SD) of adherence to uroepithelial cells caused by[a]					
Uropathogen[b]	Whole Cells	Formalin-Treated Whole Cells[c]	Acid-Treated Whole Cells[d]	Sonicated Cell Fragments (fraction A)[e]	SDS-Extracted Cell Fragments (fraction B)[f]	SDS-Extracted Acid-Treated Cell Fragments (fraction C)[g]
E. coli (MS)	75.4 ± 34.7	42.5 ± 23.4	31.2 ± 16.4	52.1 ± 5.4	36.8 ± 19.0	3.9 ± 28
E. coli (MR)	74.1 ± 36.6	71.9 ± 13.6	17.3 ± 9.8	50.9 ± 20.1	27.9 ± 12.3	4.2 ± 3.5
E. coli (encapsulated)	70.6 ± 41.2	38.1 ± 16.3	12.3 ± 6.8	74.7 ± 6.8	52.4 ± 18.1	4.8 ± 2.6
Klebsiella pneumoniae	76.9 ± 32.6	43.0 ± 19.8	24.7 ± 13.0	33.8 ± 6.1	22.7 ± 16.7	4.6 ± 3.8
Pseudomonas aeruginosa (nonmucoid)	69.6 ± 34.9	41.2 ± 16.8	29.7 ± 18.7	46.4 ± 8.2	47.5 ± 8.8	2.5 ± 1.3
P. aeruginosa (mucoid)	61.9 ± 36.9	38.1 ± 16.3	4.8 ± 0.7	42.0 ± 8.6	30.3 ± 0.14	6.3 ± 7.4

[a]Uroepithelial cells were obtained from freshly voided midstream urine samples from four healthy, premenopausal women 20–36 years old) with no past history of urinary tract infection. The urine samples were collected on day 10 of the menstrual cycle. Cells were harvested by centrifugation, washed once with phosphate-buffered saline (PBS) and suspended in PBS to a concentration of 10^5 cells/ml. Uroepithelial cells (1×10^5 cells) were preincubated with *Lactobacillus* whole cells (1×10^8 cells) or cell wall fragments (dry weight equal to the dry weight of whole cells) before being incubated with uropathogenic bacterial to test for competitive exclusion: Aliquots of [^{32}P]-labeled uropathogens and uroepithelial cells were combined, incubated at 37°C for 30 min. The cells were then filtered through 5-μm-pore-sized filters to remove free bacteria, and then washed with PBS, air dried, placed in a scintillation vial containing 15 ml scintillation fluid, and counted in a LKB 1211 Minibeta Counter. The relative number of bacteria adherent to uroepithelial cells retained on the filters was determined by subtracting the control counts (uroepithelial cells only and bacterial cells only) from the counts of the mixtures on the filters per minute. The *Lactobacillus* strain was a distal urethral isolate from a woman.
[b]The six uropathogens were isolated from urine of patients with chronic cystitis, maintained on brain–heart infusion. MS = mannose-sensitive, MR= mannose-resistant. To label two bacteria, cells were incubated in 2-ml fresh, filter-sterilized urine supplemented with 20 μCi of [^{3}H] uridine per ml, at 100 rpm, 37°C for 18 h. The bacteria were harvested by centrifugation and suspended to a concentration of 10^8 CFU/ml in PBS (pH 6.7).
[c]Whole cells of the *Lactobacillus* culture were fixed with 10% formalin in PBS at room temperature for 1 h, washed 3 times in PBS, and resuspended in PBS.
[d]Cells were incubated in 0.1 N H_2SO_4 for 3 min, neutralized with NaOH, washed 3 times with distilled water, and suspended in PBS.

[e]The *Lactobacillus* cells were harvested, washed, and suspended in PBS. For treatment A (fraction A) the cells were sonicated by ten 30-s pulses, centrifuged at 12,000*g* for 10 min, and suspended in 20 mM Tris (pH 7.4)–8 mM $MgCl_2$–1 μg of RNase A (Sigma Chemical Co.) per ml-1 μg of DNase I (Sigma) per ml, incubated at 37°C for 3 h. The sample was then washed twice and suspended in PBS.
[f]For fraction B, the fraction A sample was heated to 100°C in PBS with 1% (w/v) sodium dodecyl sulfate (SDS) for 4 h. The sample was then centrifuged 3 times at 12,000*g* for 10 min (each time), and then dialyzed against 1 l of 1% bovine serum albumin in distilled water, and resuspended in PBS.
[g]For fraction C, the cell wall fragments (fraction A) were treated with 10 ml of 0.1 N H_2SO_4, neutralized with sodium hydroxide, washed twice with distilled water, and dialyzed for 72 h in 4 l water. Fraction C (peptidoglycan) did not contain any chloroform–methanol–extractable phospholipid, did not incorporate any [$^{3}2$ P]PO_4, and did not have associated proteins.

Remarks

1. Indigenous *Lactobacillus* isolated from urethral surface of healthy woman was able to adhere to human uroepithelial cells *in vitro.* The bacterium was found to block the adherence of gram-negative uropathogenic bacteria to uroepithelial cells from women without a history of urinary infections. Competitive exclusion was most effective with whole viable cells and less effective with cell wall fragments obtained by sonication, extraction with SDS, and treatment with SDS and acid.
2. These results suggest that the normal flora of the urinary tract may be used to protect against attachment of uropathogen to the surface of uroepithelial cells.

Inhibition of Uropathogens Adherence by Lipoteichoic Acid (LTA) Isolated from *Lactobacillus* Cells

	% Inhibition (mean ± SD) of Adherence to Uroepithelial Cells Caused by		
Uropathogen[a]	LTA[b]	Deacylated LTA[c]	Reconstituted LTA Peptidoglycan[d]
E. coli (MS)	23.6 ± 4.2	8.7 ± 1.4	48.6 ± 5.2
E. coli (MS)	28.8 ± 8.7	14.6 ± 3.1	48.8 ± 6.4
E. coli (encapsulated)	27.4 ± 6.0	12.4 ± 4.2	69.5 ± 13.1
K. pneumoniae	24.4 ± 5.2	6.3 ± 0.4	40.2 ± 8.8
P. aeruginosa (nonmucoid)	24.8 ± 6.0	4.8 ± 2.0	44.0 ± 6.6
P. aeruginosa (mucoid)	8.7 ± 1.4	0.0	31.2 ± 2.8

[a]Uroepthelial cells (1×10^5) in 1 ml PBS were preincubated with *Lactobacillus* LTA, deacylated LTA, or reconstituted LTA–peptidoglycan (dry weight equal to the dry weight of 10^8 whole cells) before being incubated with uropathogenic bacteria to test for competitive exclusion.
[b]*Lactobacillus* cells were grown overnight in filter-sterilized urine supplemented with 10 μCi of [^{32}P] PO_4 per milliliter. The cells were harvested and the LTA was extracted by boiling the cells for 1 h in 100% phenol. The LTA-containing cell wall fragment was then washed in PBS by centrifugation, suspended in PBS, and dialyzed overnight against water. The LTA was protein free and did not contain chloroform–methanol–extractable phospholipid. The property of competitive exclusion was measured as percent inhibition of the adherence of uropathogens, by scintillation counting of the amount of ^{32}P bound to uroepithelial cells.

[c]Sample of LTA was incubated in 0.1 N ammonium hydroxide overnight to deacylate the LTA. The sample was then neutralized and suspended in water.
[d]Excess amounts of purified, nonlabeled LTA were incubated with peptidoglycan (fraction C of the SDS extract) at 37°C for 1 h. The excess LTA was then removed by centrifugation. The reconstituted cell wall fragments were dialyzed and suspended in PBS before used.

Remarks The study suggested that lipoteichoic acid was responsible for the adherence of *Lactobacillus* cells to uroepithelial cells but that steric hindrance was the major factor in preventing the adherence of uropathogens. This conclusion was supported by blockage studies with reconstituted lipoteichoic acid–peptidoglycan, which was more effective at blocking adherence than lipoteichoic acid or peptidoglycan alone.

REFERENCES

1. Barrow, P. A., Brooker, B. E., Fuller, R., and Newport, M. J. The attachment of bacteria to the gastric epithelium of the pig and its importance in the microecology of the intestine. *J. Appl. Bacteriol.* **48:** 147–154, 1980.
2. Chan, R. C. Y., Reid, G., Irvin, R. T., Bruce, A. W., and Costerton, J. W. Competitive exclusion of uropathogens from human uroepithelial cells by *Lactobacillus* whole cells and cell wall fragments. *Infect. Immun.* **47**(1): 84–89, 1985.
3. Mead, G. C., Barrow, P. A., Hinton, M. H., Humbert, F., Impey, C. S., Lahellec, C., Mulder, R. W. A. W., Stavric, S., and Stern, N. J. Recommended assay for treatment of chicks to prevent *Salmonella* colonization by competitive exclusion. *J. Food Prot.* **52:** 500–502. 1989.
4. Stavric, S., Gleeson, T. M., Blanchfield, B., and Pivnick, H. Role of adhering microflora in competitive exclusion of *Salmonella* from young chicks. *J. Food Prot.* **50**(11), 928–932, 1987.
5. Watkins, B. A., Miller, B. F., and Neil, D. H. *In vivo* inhibitory effects of *Lactobacillus acidophilus* against pathogenic *Escherichia coli* in gnotobiotic chicks. *Poult. Sci.* **61:** 1298–1308, 1982.

5.2 PRODUCTION OF ANTIMICROBIAL SUBSTANCES

All lactic acid bacteria produce organic acid. Acetic acid (pK_a = 4.75) has been shown a more effective inhibitor of microorganisms than lactic acid (pK_a = 3.08) (10). Acetic acid is produced by *Leuconostoc citrovorum* and has been shown to inhibit *Salmonella gallinarium* (32) and *P. fragi* (26).

Lactobacillus species, *Lactococcus lactis,* and *Leuconostoc cremoris* produce hydrogen peroxide when transferred from anaerobic to aerobic condition (5) and incubated at low temperature (7). The hydrogen peroxide produced was found to inhibit the growth of *S. aureus* (7), various psychotrophic microorganisms (21) and Pseudomonas (27). Hydrogen peroxide is also involved in the lactoperoxidase system of raw milk (29).

Other anitmicrobial agents produced by probiotic organisms are strain specific and are listed in the following tables.

Bacteria	*L. acidophilus* N2	*L. acidophilus* 11088 (NCK88)	*L. acidophilus*	*L. acidophilus*	*L. bulgaricus* DDS14
Antibacterial agent	Lactacin B (Bacteriocin)	Lactacin F (Bacteriocin)	Lactocidin (Bacteriocin-like)	Acidolin	Bulgarican
Molecular weight (daltons)	6200	2500		200	
Optimum pH	5		Neutral		<4.5
Thermostability	100°C, 60 min	121°C, 15 min			
Sensitive to proteinaceous enzymes	Proteinase K	Trypsin, proteinase K, subtilisin, ficin			
Structure		56 amino acids			Low MW peptide
Gene control	Chromosomal linked	Transient plasmid determinants			
Bactericidal mechanism					
Bactericidal effect					
Gram positive (+)					
Staphylococcus				–	+
Streptococcus			+		
Lactococcus	+		+		
Mycobacterium					+
Micrococcus					
Bacillus			+	+	+
Leuconostoc					
Clostridium	–				
Listeria					
Bifidobacterium					
Corynebacterium					
Lactobacillus	+	+	+	–	
Pediococcus					
Gram negative (–)					
Campylobacter					
Escherichia			+	+	+
Pseudomonas					
Salmonella			+		
Vibrio					
Ref.	3	25, 25a	37	15	28

Bacteria	*L. rhamnosus* GG	*L. reuterii*	*L. helveticus, L. fermenti* 466	*L. helveticus*	*L.plantarum*
Antibacterial agent	Microcin	Reuterin	Lactocin 27	Helveticin J	Lactolin
Molecular weight (Daltons)	<1000		12,400	37,000	100,000

(continued)

Bacteria	*L. rhamnosus* GG	*L. reuterii*	*L. helveticus*, *L. fermenti* 466	*L. helveticus*	*L.plantarum*
Optimum pH	3–5				5
Thermostability	121°C, 15 min		100°C, 60 min	<100°C, 30 min	<121°C, 15 min
Sensitive to proteinaceous enzymes			Trypsin, pronase	Subtilisin, pronase, pepsin, trypsin, proteinase K, ficin	Lipase, α-amylase, trypsin, pepsin
Structure	Low MW peptide		Glycoprotein	Polypeptide	
Gene control	Plasmid mediated		No evidence of plasmid control	Chromosomal control	
Bactericidal mechanism			Termination of protein synthesis		
Bactericidal effect					
Gram positive (+)					
Staphylococcus	+	+			+
Streptococcus	+				
Lactococcus					
Mycobacterium	+				
Micrococcus					
Bacillus	+				
Leuconostoc					+
Clostridium	+	+			+
Listeria	+	+		–	
Bifidobacterium	+				
Corynebacterium					
Lactobacillus	–		+	+	+
Pediococcus					+
Gram negative (–)					
Campylobacter					
Escherichia	+	+			
Pseudomonas					
Salmonella	+	+			
Vibrio					
Ref.	2, 31	6	35, 36	16, 20	1, 38

Bacteria	*L. plantarum* C-11	*L. sake* 706	*Lactococcus lactis* ssp. *lactis* DL16 *Lactococcus lactis* ssp *lactis* SIK83 *Lactococcus lactis* ssp. *lactis* 11454	*Streptococcus cremoris* 202 *L. lactis* ssp. *cremoris* *L. lactis* ssp. *lactis diacetylactis* *L. lactis* ssp. *lactis*
Antibacterial agent	Plantaricin A	Sakacin A	Nisin (Bacteriocin)	Lactostrepcins 1, 2, 3, 4, 5
Molecular weight (daltons)	<8000		3500	6000–20,000
Optimum pH	4.0–6.5		2.5–5.5	4.6–5.0

(*continued*)

Bacteria	*L. plantarum* C-11	*L. sake* 706	*Lactococcus lactis* ssp. *lactis* DL16 *Lactococcus lactis* ssp *lactis* SIK83 *Lactococcus lactis* ssp. *lactis* 11454	*Streptococcus cremoris* 202 *L. lactis* ssp. *cremoris* *L. lactis* ssp. *lactis diacetylactis* *L. lactis* ssp. *lactis*
Thermostability	100°C, 30 min		115.6°C, pH 2	100°C, 10 min/121°C 15 min
Sensitive to proteinaceous enzymes			Pancreatin, α-chymotrysin	Trypsin, α-chymotrypsin, pronase, phospholipase
Structure	No evidence of plasmid control	Plasmid DNA	34 amino acids, 5 thioether bonds	
Gene control			28–30 MDa plasmid/ chromosomal	Chromosomal control
Bactericidal mechanism			Surface-active cationic detergent	Membrane disruption. Inhibit: RNA, DNA, protein synthesis
Bactericidal effect				
Gram positive (+)				
Staphylococcus			+	
Streptococcus				+
Lactococcus			+	
Mycobacterium			+	
Micrococcus			+	
Bacillus				+
Leuconostoc	+		–	
Clostridium		–	+	
Listeria	–	+	+	
Bifidobacterium				
Corynebacterium			+	
Lactobacillus	+		+	+
Pediococcus	+			
Gram negative (–)				
Campylobacter				
Escherichia		–	–	
Pseudomonas			–	+
Salmonella		–	–	
Vibrio			+	
Ref.	6a	24, 30	17, 19, 23, 24, 33, 34	8, 9, 22, 39

Bacteria	*Streptococcus cremoris* 346	*Pediococcus pentosaceus* 4321/43200/ L7230/FBB61/ FBB63	*Pediococcus acidilactici* PC/B5627/P02/ PAC1.0	*Pediococcus acidilactici* H
Antibacterial agent	Diplococcin (Bacteriocin)	Pediocin A (Bacteriocin)	Pediocin PA-1 (Bacteriocin)	Pediocin Ac H (Bacteriocin)
Molecular weight (daltons)	5300		16,500	2700

(*continued*)

Bacteria	*Streptococcus cremoris* 346	*Pediococcus pentosaceus* 4321/43200/ L7230/FBB61/ FBB63	*Pediococcus acidilactici* PC/B5627/P02/ PAC1.0	*Pediococcus acidilactici* H
Optimum pH			5.5–7.0	2.5–9.0
Thermostability		100°C, 60 min	100°C, 10 min	121°C, 15 min
Sensitive to proteinaceous enzymes	Trypsin, pronase, α-chymotrypsin	Pronase	Pepsin, α-chymotrypsin, papain, pronase	Trypsin, chymotrypsin, ticin, papain, proteases IV, XIV, XXIV, K
Structure	Lanthionine, β-methyl lanthionine, dehydroalanine, dehydrobutyrine similar to Nisin, but lacks S-containing amino acid.			
Gene control	Plasmid associate	10.5 MDa plasmid	6.2 MDa plasmid	
Bactericidal mechanism				Destabilizes membrane functions
Bactericidal effect				
Gram positive (+)				
Staphylococcus		+	+	
Streptococcus		+	+	+
Lactococcus			–	
Mycobacterium				
Micrococcus		+		
Bacillus	–	+	+	
Leuconostoc		+	+	
Clostridium	–	+		–
Listeria				+
Bifidobacterium			+	
Corynebacterium				
Lactobacillus		+	+	
Pediococcus	+	+	+	
Gram negative (–)				
Campylobacter		–		+
Escherichia		–		
Pseudomonas				
Salmonella				
Vibrio		–		
Ref.	8	11, 14, 33	13, 18	4, 24

REFERENCES

1. Andersson, R. E. Inhibition of *Staphylococcus aureus* and spheroplasts of gram-negative bacteria by an antagonistic compound produced by a strain of *Lactobacillus plantarum*. *Int. J. Food Microbiol.* **3**: 149–160, 1986.
2. Asensio, C., Perez-Diaz, J. C., Martinez, M. C., and Baquero, F. A new family of low

molecular weight antibiotics from enterobacteria. *Biochem. Biophys. Res. Commun.* **67:** 7–14, 1976.

3. Barefoot, S. F., and Klaenhammer, T. R. Detection and activity of Lactacin B, a bacteriocin produced by *Lactobacillus acidophilus. Appl. Environ. Microbiol.* **45:** 1808–1815, 1983.
4. Bhunia, A. K., Johnson, M. C., and Ray, B. Purification, characterization and antimicrobial spectrum of a bacteriocin produced by *Pediococcus acidilactici. J. Appl. Bacteriol.* **65:** 261–268, 1988.
5. Collins, E. B., and Aramaki, K. Production of hydrogen peroxide by *Lactobacillus acidophilus. J. Dairy Sci.* **63:** 353–357, 1980.
6. Daeschel, M. A. Antimicrobial substances from lactic acid bacteria for use as food preservatives. *Food Technol.* **43:** 164, 1989.

6a. Daeschel, M. A., McKenny, M. C., and McDonald, L. C. Bacteriocidal activity of *Lactobacillus plantarum* C-11. *Food Microbiology* **7:** 91–98, 1990.

7. Dahiya, R. S., and Speck, M. L. Hydrogen peroxide formation by Lactobacilli and its effect on *Staphylococcus aureus. J. Dairy Sci.* **51**(10): 1568–1572, 1968.
8. Davey, G. P., and Richardson B. C. Purification and some properties of diplococcin from *Streptococcus cremoris* 346. *Appl. Environ. Microbiol.* **41:** 84, 1981.
9. Dobrzanski, W. T., Bardowski, J., Kozak, W., and Zajdel, J. Lactostrepcins: Bacteriocins of lactic streptococci. In *Microbiology-1982* (D. Schlessinger, ed.), p. 225. American Society for Microbiology, Washington, DC, 1982.
10. Doores, S. pH control agents and acidulants. In *Food Additives* (A. L. Branen, P. M. Davidson, and S. Salminen, eds.), pp. 477–510. Dekker, New York, 1990.
11. Fleming, H. P., Etchells, J. L., and Costilow, R. N. Microbial inhibition by an isolate of *Pediococcus* from cucumber brines. *Appl. Microbiol.* **30:** 1040, 1975.
12. Gilliland, S. E. Role of starter culture bacteria in food preservation. In *Bacterial Starter Cultures for Food* (S. E. Gilliland, ed.), p. 175. CRC Press, Boca Raton, FL, 1985.
13. Gonzales, C. F., and Kunka, B. S. Plasmid-associated bacteriocin production and sucrose fermentation in *Pediococcus acidilactici. Appl. Environ. Microbiol.* **53:** 2434–2538, 1987.
14. Graham, D. C., and McKay, L. L. Plasmid, DNA in strains of *Pediococcus cerevisiae* and *Pediococcus pentosaceus. Appl. Envron. Microbiol.* **50:** 532, 1985.
15. Hamdan, I. Y., and Mikolajcik, E. M. Acidolin: an antibiotic produced by *Lactobacillus acidophilus. J. Antibiot.* **27:** 631–636, 1974.
16. Harris, L. J., Daeschel, M. A., Stiles M. E., and Klaenhammer, T. R. Antimicrobial activity of lactic acid bacteria against *Listeria monocytogenes. J. Food Prot.* **52:** 384–387, 1989.
17. Harris, L. J., Fleming, H. P. and Klaenhammer, T. R. Sensitivity and resistance of *Listeria monocytogenes* ATCC 19115, Scott A., and UAL 500 to nisin. *J. Food Prot.* **54:** 836–840, 1991.
18. Hoover, D. G., Walsh, P. M., Kolaetis, K. M., and Daly, M. M. A bacteriocin produced by *Pediococcus* species associated with a 5.5-mega dalton plasmid. *J. Food Prot.* **51:** 29–31, 1988.
19. Hurst, A., and Hoover, D. G. Nisin. In *Antimicrobials in Foods* (P. M. Davidson and A. L. Branen, eds.), 2nd ed., p. 369. Dekker, New York, 1993.
20. Joerger, M. C., and Klaenhammer, T. R. Characterization and purification of helveticin J and evidence for a chromosomally determined bacteriocin produced by *Lactobacillus helveticus* 481. *J. Bacteriol.* **167:** 439–446, 1986.

21. Juffs, H. S., and Babel, F. J. Inhibition of psychrotrophic bacterin by lactic cultures in milk stored at low temperature. *J. Dairy Sci.* **58:** 1612–1619, 1975.

22. Kozak, W., Bardowski, J., and Dobrzanski W. T. Lactostrepcins-acid bacteriocins produced by lactic streptococci. *J. Dairy Res.* **45:** 247, 1978.

23. Kozak, W., Raj-chert-Trzpil, M., and Dobrzanski, W. T. The effect of proflavin, ethidium bromide and an elevated temperature on the appearance of nisin-negative clones in nisin-producing strains of *Streptococcus lactis. J. Gen. Microbiol.* **83:** 295–302, 1974.

24. Motlagh, A. M., Johnson, M. C., and Ray, B. Viability loss of food borne pathogens by starter culture metabolites. *J. Food Prot.* **54:** 873–878, 1991.

25. Muriana, P. M., and Klaenhammer, T. R. Conjugal transfer of plasmid-encoded determinnants for bacteriocin production and immunity in *Lactobacillus acidophilus* 88. *Appl. Environ. Microbiol.* **53:** 553, 1987.

25a. Muriana, P. M., and Klaenhammer, T. R. Purification and partial characterization of lactacin F and bacteriocin produced by *Lactobacillus acidophilus* 11088. *Appl. Environ. Microbiol.* **57**(1): 114–121, 1991.

26. Pinheiro, A. J. R., Liska, B. J., and Parmelee, C. E. Properties of substances inhibitory to *Pseudomonas fragi* produced by *Streptococcus citrovorus* and *S. diacetylactis. J. Dairy Sci.* **51:** 183–187, 1968.

27. Price, R. J., and Lee, J. S. Inhibition of *Pseudomonas* species by hydrogen peroxide producing lactobacilli. *J. Milk Food Technol.* **33:** 13, 1970.

28. Reddy, N. S., and Ranganathan, B. Preliminary studies on antimicrobial activity of *Streptococcus lactis* subsp. *diacetylactis. J. Food Prot.* **46:** 222–225, 1983.

29. Reiter, B., and Harnulv, B. G. Lactoperoxidase antibacterial system; natural occurrence, biological functions and practical applications. *J. Food Prot.* **47:** 724–732, 1984.

30. Schillinger, U., and Lucke, F. K. Antibacterial activity of *Lactobacillus sake* isolated from meat. *Appl. Environ. Microbiol.* **55:** 1901–1906, 1989.

31. Silva, M., Jacobus, N. V., Deneke, C., and Gorbach, S. L. Antimicrobial substance from a human *Lactobacillus* strain. *Antimicrob. Agents Chemother.* **31**(8): 1231–1233, 1987.

32. Sorrels, K. M., and Speck, M. L. Inhibition of *Salmonella gallinarum* by culture filtrates of *Leuconostoc citrovorum. J. Dairy Sci.* **53:** 239, 1970.

33. Spelhaug, S. R., and Harlander, S. K. Inhibition of foodborne bacterial pathogens by bacterocins from *Lactococcus lactis* and *Pediococcus pentosaceous. J. Food Prot.* **52:** 856–862, 1989.

34. Steen, M. T., Chung, Y. J., and Hansen, J. N. Characterization of the nisin gene as part of a polycistronic operon in the chromosome of *Lactococcus lactis* ATCC 11454. *Appl. Environ. Microbiol.* **57:** 1181–1188, 1991.

35. Upreti, G. C., and Hinsdill, R. D. Isolation and characterization of a bacteriocin from a homofermentative *Lactobacillus. Antimicrob. Agents Chemother.* **4:** 487, 1973.

36. Upreti, G. C., and Hinsdill, R. D. Production and mode of action of lactocin 27: Bacteriocin from homofermentative *Lactobacillus. Antimicrob. Agents Chemother.* **7**: 139, 1975.

37. Vincent, J. G., Veomett, R. C., and Riley, R. F. Antibacterial activity associated with *Lactobacillus acidophilus. J. Bacteriol.* **78:** 477–484, 1959.

38. West, C. A., and Warner, P. J. Plantacin, B, a bacteriocin produced by *Lactobacillus plantarum* NCDO 1193. *FEMS Microbiol. Lett.* **49:** 163, 1988.

39. Zajdel, J. K., Ceglowski, P., and Dobrzanski, W. T. Mechanism of action of lactostrepcin 5, a bacteriocin produced by *Streptococcus cremoris* 202. *Appl. Environ. Microbiol.* **49:** 969, 1985.

5.3 MODULATION OF IMMUNE RESPONSES

Target	Animal	Route	Viability	Probiotics	Effect	Ref.
Host defense	BALB/c, C3H/HE, C57BL/6, Athymic nu/nu, Athymic nu/+ mice	Subcutaneous	Heat killed	*L. casei* strain Shirota	Enhance resistance to systemic *Listeria monocytogenes* infection.	5
	Swiss albino mice	Oral	Live	*L. casei*	Prevent enteric infection by stimulating sIgA production.	6
B-lymphocyte	C57BL/6	*In vitro*	Heat killed	*L. plantarum* ATCC 14917, *L. fermentum* YIT 0159	Stimulate activity of B-lymphocyte population of splenocytes.	10
	Swiss albino mice	Oral	Live	*L. casei* *L. acidophilus* Fermented milk	Increase IgM forming cells in spleen and circulating antibody against SRBC inocu-lated intraperitoneally. Produce neither hepatomegaly nor splenomegaly.	7
	Swiss albino mice	Oral	Live	*L. casei,* *L. acidophilus* yogurt	Increase IgA forming cells in small intestine. *L. casei* increase secretary IgA against *S. typhimurium*. *L. acidophilus* decrease lymphocytes & IgA producer in intestine after 7 days feeding.	6
	BALB/c mice Peyer's patch cells	*In vitro*	Formalin killed or cell wall fraction	*Bifidobac-terium breve* strain Yakult	Cell wall fraction activated macrophage-like cells that secreted a soluble factor that enhanced B cell proliferation.	11
	Mice	Oral	Live	*B. breve* strain Yakult	Increased antigen specific IgA titer in milk and intestine of dams.	12
Phagocytes	ICR, C57BL/6 mice	Intraperitoneal	Heat killed	*L. casei* strain Shirota	Increase activity of macrophage, reticuloendotherial cells, lysosomal enzymes, and cytostatic activity of peritoneal exudate cells on leukemic cells.	2
	Swiss albino mice	Oral	Live	*L. casei, L. acidophilus*	Increase *in vitro* and *in vivo* phagocytic activity.	7

(*continued*)

Target	Animal	Route	Viability	Probiotics	Effect	Ref.
Phagocytes (*cont.*)	Healthy human	Oral	Live	*L. acidophilus* La1 *B. bifidum* Bb12	Enhance phagocytic activity of granulocyte and monocyte.	9
T-lymphocyte	C57BL/6CrSlc mice	Oral	Peptido-glycan extract	*B. thermophilum* P2-91	Enhanced cytotoxic activity of NK cells in spleen and mesenteric lymph node.	8
	BALB/c, C3H/He, CBA/N, C57BL/6 mice	Intravenous	Heat killed	*L. casei* strain Shirota	Augment NK activity of spleen in normal and tumor bearing mice	3
Hematopoiesis	BALB/c mice	Subcutaneous	Heat killed	*L. casei* strain Shirota	Increase monocyte/macrophage in spleen. Increase macrophage progenitor cells in spleen and femur.	13
Allergic reaction	BALB/c mice	Oral	Live	*L. casei* strain Shirota	Inhibited IgE production. Enhanced Th1 cell-associated cytokines production, suppressed Th2 cell-associated cytokines production.	4

5.3.1 Effect on Host Defense Mechanism

Enhancement of Resistance to Systemic *Listeria monocytogenes* Infection (5)

Animal[a]	Organ	Control[b] (Log_{10} viable bact. per organ)	Fed with LC 9018[c] (Log_{10} viable bact. per organ)	p
Male BALB/c	Liver	7.93 ± 0.54	6.48 ± 0.50	< 0.01
	Spleen	7.63 ± 0.10	7.05 ± 0.11	< 0.01
Male C3H/He	Liver	7.10 ± 0.15	5.40 ± 0.20	< 0.01
	Spleen	6.34 ± 0.06	5.35 ± 0.10	< 0.01
Male ICR	Liver	6.55 ± 0.28	5.29 ± 0.63	< 0.01
	Spleen	6.25 ± 0.23	5.74 ± 0.45	< 0.01
Female C57BL/6	Spleen	6.28 ± 0.10	5.44 ± 0.17	< 0.01
Athymic nu/nu	Liver	7.28 ± 0.04	6.67 ± 0.27	< 0.01
Athymic nu/+	Liver	7.63 ± 0.22	6.04 ± 0.19	< 0.01

[a]Mice were 7–11 weeks old.

[b]Mice were challenged intravenously with $1–5 \times 10^6$ CFU of *Listeria monocytogenes* and the number of *Listeria monocytogenes* in the liver and spleen was counted 24 h later. All untreated mice died by day 3 after the challenge.

[c]A 0.5-mg amount of lyophilized heat-killed *L. casei* strain Shirota (LC 9018) suspended in PBS, was administered subcutaneously near the inguino-abdomen or axilla, 5 days before the mice were challenged intravenously with $1–5 \times 10^6$ CFU of *Listeria monocytogene.* The mice were dissected and the

number of *Listeria monocytogenes* in liver and spleen counted 24 h later. The survival rate after the challenge peaked at 5–6 days after administration of LC 9018 (83% survival).

Remarks Subcutaneous administration of heat-killed *L. casei* strain Shirota into mice enhanced the host resistance to systemic infection with *Listeria monocytogenes* in normal and athymic nude mice, and the resistance was partly T-lymphocyte dependent. Growth of *Listeria monocytogenes* in the liver and spleen was strongly inhibited five days after the administration of LC 9018.

Prevention of *Salmonella typhimurium* Infection (6)

	Day after Challenged[a]	Nourished[b]		Malnourished[c]		
		Control	Fed with *L. casei*[d]	Control	Fed with Milk	Fed with *L. casei*[d]
Bacteria no.[e]	2	1.0	0.0	3.9	1.1	0.0
	5	3.9	0.8	4.1	4.0	4.0
	7	3.1	0.5	3.8	3.7	4.0
IgA[f]	2	0.8	2.5	0.5	0.6	0.4
	5	0.6	2.3	0.4	0.6	0.8
	7	1.0	1.3	0.4	0.5	0.5
IgM[g]	2	3.0	3.0	1.3	0.8	0.5
	5	2.9	3.6	1.1	2.3	2.3
	7	2.2	3.1	1.0	2.2	2.2
IgA cells[h]	0	100	120	55	98	110

[a]Mice were challenged orally with 10^6 *Salmonella typhimurium* at 20 times median lethal dose.
[b]Well-nourished Swiss albino mice.
[c]Fed with fermented milk containing 1.2×10^9 *L. casei*/day/mouse.
[d]Mice malnourished by protein deprivation at weaning.
[e]Log_{10} number of viable *Salmonella typhimurium* isolated from liver and spleen in the colony count.
[f]Concentration of secretory IgA in intestinal fluid was measured by ELISA and expressed as OD_{493nm}/100 ml.
[g]Concentration of IgM in intestinal fluid was measured by ELISA and expressed as OD_{493nm}/100 ml.
[h]Number of IgA producing cells was determined by immunofluorescent test on histological slices of the small intestine.

Remarks Orally administered *L. casei* could prevent enteric infection by stimulating sIgA synthesis. The number of IgA producing cells and IgM in the intestine of *L. casei* treated mice were increased. *L. casei* could be used as adjuvant in oral vaccine. In the malnourished mice, feeding with fermented milk containing *L. casei* for 2 days could not enhance the immunoactivity of the animal above that fed with just milk for the same period of time.

5.3.2 Effect on B-lymphocyte

Effect on Activity of B-lymphocyte Population of Splenocytes (10)

Bacterial Strain	Dose[a] (μg/ml)	Mitogenicity[b] (cmp)	p
L. plantarum ATCC14917	100	575 ± 172	$0.02 < p < 0.05$
	100+ATS[c]	724 ± 21	$p < .001$
	control	206 ± 8	
L. fermentum YIT 0159	100	1617 ± 25	p < 0.001
	50	1463 ± 90	$p < 0.001$
	10	1429 ± 21	$p < 0.001$
	100+ATS	957 ± 46	
	control	361 ± 41	
LPS from *S. typhimurium*	10	2110 ± 802	
	10+ATS	5657 ± 932	
Concanavalin A	1	8546 ± 535	
	1+ATS	187 ± 4	

[a]Six-to-twelve-week-old C57BL/6 male mice (Shizuoko Agricultural Cooperative for Laboratory Animals, Hamamatsu, Japan) were used throughout the experiments.
[b]Mitogenicity assay: (a) Spleen was removed and teased with forceps in cold Eagle's minimal essential medium (Nissui Seiyaku, Tokyo). After brief sedimentation, cells in supernatant were collected, washed with Eagle's medium, and suspended in 0.83% NH_4Cl buffer for 1–2 min until erythrocytes were lysed. (b) Residual lymphocytes were washed with Eagle's medium and their viability was determined by trypan blue dye exclusion. The cell concentration was adjusted to 8×10^6/ml by suspending cells in RPMI-1640 medium (GIBCO, USA) supplemented with 10% fetal calf serum (GIBCO, USA).(c) A 0.25-ml cell suspension and 0.15 ml (dose as indicated in table) of heat-killed lactobacilli were placed in a culture tube (16 × 125 mm) and the final volume was adjusted to 3 ml with RPMI medium. The suspension was incubated at 37°C in a humidified CO_2 (5% in air) incubator for 48 h. (d) The culture was pulsed with 1 μCi ^{3}H-thymidine (Radiochemical Center, England) and incubated for 16 h. (e) Cells were harvested on a glass fiber filter (Whatman, GF/C), washed with Hank's solution and 5% trichloroacetic acid and dried. Radioactivity taken up by cells was measured with a scintillation counter (Aloka, LCS-661, Tokyo). The results were expressed as mean of counts per min (cpm) per culture.
[c]ATS, rabbit antithymocyte serum.

Remarks *L. fermentum* and *L. plantarum* stimulated the incorporation of ^{3}H-thymidine into splenocytes, even in the presence of rabbit antithymocyte serum and complement, designed to kill T- lymphocytes. Lipopolysaccharide (LPS) from *S. typhimurium* and concanavalin A served as the control. These results indicate that *L. fermentum* and *L. plantarum* act on B-lymphocyte population of splenocytes.

Effect on Activity of B-lymphocytes in Peyer's Patch (11)

		Proliferation Rate (cpm)[a]		
Animal	Mitogen	Peyer's Patch Cells[b]	B Cell-Enriched Fraction[c]	B-Cell Fraction[d]
Male[e] BALB/c mice	Control	401 ± 44	69 ± 13	35 ± 5
	+ *B. breve*[f] strain Yakult	2376 ± 261[g]	815 ± 69[h]	49 ± 3
	+ *B. breve* strain Yakult + polymyxin B	6720 ± 290[h]	—	

(continued)

Effect on Activity of B-lymphocytes in Peyer's Patch (11) (*cont.*)

Animal	Mitogen	Proliferation Rate (cpm)[a]		
		Peyer's Patch Cells[b]	B Cell-Enriched Fraction[c]	B-Cell Fraction[d]
	+ *B. breve* strain Yakult cell wall[i]	4920 ± 360[h]	—	
	+ LPS[j]	9228 ± 381[h]	3643 ± 139[h]	
	+ LPS + polymyxin B	1728 ± 224[g]	—	
Nude[e] (nu/nu) mice	Control	4529 ± 585		
	+ *B. breve* strain Yakult	9119 ± 432[g]		
	+ LPS	19588 ± 1096[h]		

[a]The cell suspensions were pulsed with ^{3}H-thymidine (0.5 μCi, 2.0 Ci/mmol, New England Nuclear Corp. Boston, MA). Its uptake over 18 h was counted with liquid scintillation counter (Packard Instrument Co. Inc. Rockville, MD) and expressed as count per minute (cpm).
[b]Cells of Peyer's patch (PP), a gut-associated lymphoid tissue were prepared (3). For measurement of proliferation, 5×10^5 PP cells in 0.2 ml RPMI 1640 medium supplemented with 5% FBS were cultured in 96-well Nunc trays with 1×10^7 *B. breve* strain Yakult cells, 10 μg *B. breve* cell wall or 50 μg/ml (LPS) in a 37°C, 5% CO_2 incubator.
[c]Obtained from PP cells by panning on plastic dishes (15 × 100 mm, Falcon Labware, Oxnard, CA) coated with goat anti-mouse Ig (Capped Laboratories, Cochranville, PA). The adherent B-cell-enriched fraction was then removed by vigorous pipetting.
[d]B cell fraction containing >93% immunoglobulin$^+$ cells and <1% plastic-adherent cells (PAC) was prepared by exclusion of PAC from B-cell-enriched fraction. PAC fraction was isolated by incubation in uncoated plastic dishes at 37° C for 2 h.
[e]The mice were 7–12 weeks old. The parental stocks were obtained from SLC Inc., Japan.
[f]*B. breve* strain Yakult was isolated from feces of healthy infants in 1968 by R. Tanaka (Yakult Central Institute for Microbiological Research, Tokyo, Japan). The cells were killed with 0.3% formalin at 37°C overnight and washed before use.
[g]Significant difference from control: $p < 0.01$.
[h]Significant difference from control: $p < 0.001$.
[i]Cell walls (cw) of *B. breve* strain Yakult were prepared as described by Araki (1).
[j]*E. coli* lipopolysaccharide (LPS) (055: B5, Difco Laboratories, Detroit, MI) is a mitogen, at dose of 50 μg/ml.

Remarks Addition of *B. breve* strain Yakult accelerated proliferation of Peyer's patch cells from normal and nude mice in *in vitro* studies. Enhancement was not found in B cell fraction in which plastic-adherent cells (macrophage-like cells) were excluded, suggesting that *B. breve* strain Yakult activated macrophage-like cells, and these cells secreted a soluble factor that enhanced proliferation of B cells.

The mitogenicity of *B. breve* strain Yakult was not inhibited by polymyxin B; thus it was not due to possible contamination of LPS during the experiment. It was the cell wall (cw) fraction of *B. breve* strain Yakult that induced mitogenicity.

Effect on IgM Forming Cells in Spleen (7)

	Treatment[a]	Day[b]	Activity
Plaque-forming cells[c]	Nonfermented milk	5	277.8 ± 47.2
	Nonfermented milk supernatant[d]	5	250.0 ± 60.0
	L. casei fermented milk	5	1500.1 ± 52.8

(*continued*)

Effect on IgM Forming Cells in Spleen (7) (*cont.*)

	Treatment[a]	Day[b]	Activity
	L. casei fermented milk supernatant[d]	5	656.0 ± 200.0
	L. acidophilus fermented milk	5	1475.1 ± 77.8
	L. acidophilus ferm. milk supernatant	5	462.0 ± 81.0
	L. casei + *L. acidophilus* ferm. milk	5	1080.6 ± 52.8
	L. casei + *L. acidophilus* milk supern.	5	497.0 ± 102.0
Antibody[e]	Nonfermented milk	10	206.0 ± 64.0
	Nonfermented milk supernatant[d]	10	256.0 ± 140.0
	L. casei fermented milk	10	1526.0 ± 72.0
	L. casei fermented milk supernatant[d]	10	576.0 ± 322.0
	L. acidophilus fermented milk	10	1356.0 ± 62.0
	L. acidophilus ferm. milk supernatant	10	760.0 ± 150.0
	L. casei + *L. acidophilus* ferm. milk	10	1796.0 ± 54.0
	L. casei + *L. acidophilus* milk supern.	10	358.0 ± 140.0

Reproduced with the permission of Blackwell Science Ltd.

[a]Swiss albino mice, 25–30 g (from Centro de Referencia para Lactobacilos, Chacabuco, Argentina) were fed with milk or fermented milk containing 100 μg cell protein (2.4 × 10^9 cells) or milk supernatant daily. Both *L. casei* and *L. acidophilus* were isolated from human feces. Each data point represented the average of 8–10 mice.

[b]Day when the maximum activity was measured.

[c]Sheep red blood cells inoculated intraperitoneally on the eighth day of feeding experiment, and the mice were killed on various days following sheep red blood cells (SRBC) inoculation. The spleen was disintegrated and diluted in Gey's solution to a 10^6 cell suspension and then mixed with 0.1 ml 20% SRBC in molten 1.5% agarose A 37 and poured into a petri dish. Incubated for 2 h at 37°C in 5% CO_2 atmosphere. Then, 10% guinea pig serum diluted in Gey's solution was added and spread over the surface. The plate was incubated for a further 40–50 min at 37°C, and hemolysis in gel was counted directly. Result was expressed as plaque-forming cell number/10^6 spleen cells.

[d]Supernatant of milk or fermented milk, centrifuged at 10,000*g* for 10 minutes.

[e]Sheep red blood cells were inoculated on the sixth, seventh, and eighth days, and mice bled from retro-orbital venous plexus on various days post-SRBC inoculation. The serum diluted and the antibody titer determined against 1% SRBC with heamagglutination reaction.

Remarks Mice fed with fermented milk containing *L. casei* and/or *L. acidophilus* showed an increase in both the number of IgM forming cells in the spleen and the circulating antibody against sheep red blood cells inoculated intraperitoneally. The activation of the immune response was partly due to substances released by these organisms into the culture supernatant, during the fermentation process. It was also noted that the feeding with fermented milk produced neither hepatomegaly nor splenomegaly.

Effect on IgA-Forming Cells in Small Intestine (6)

	No. of Mice	Treatment[a]	Days of Treatment	Activity	p
IgA-producing cells[b]	3	Control		67.9	
	3	*L. casei*	5	187.0 ± 8.3	< 0.01
	3		7	138.9 ± 7.9	< 0.05

(*continued*)

Effect on IgA-Forming Cells in Small Intestine (6) (*cont.*)

	No. of Mice	Treatment[a]	Days of Treatment	Activity	p
	3	*L. acidophilus*	5	108.0 ± 8.0	< 0.05
	3		7	10.7	
	3	Yogurt	5	111.2 ± 5.8	< 0.05
	3		7	150.0 ± 7.1	< 0.05
IgA[c]	5	Control		1.5 ± 0.1	
	5	*L. casei*	2	2.6 ± 0.2	< 0.05
	5	*L. acidophilus*	2	1.8 ± 0.1	
	5	Yogurt	5	1.7 ± 0.2	

[a]Swiss albino mice were fed with milk or fermented milk containing the different lactic acid bacteria at 1.2×10^9 cells/d/mouse. *L. casei* and *L. acidophilus* were isolated from human feces. Yogurt contained *L. delbrueckii* ssp. *bulgaricus* and *S. thermophilus*.
[b]The number of cells producing IgA in the small intestine was determined by immunofluorescence test on histological slice. The unit was cells producing IgA/10 pilli.
[c]The mice were fed with lactobacilli for 2 days and with yogurt for 7 days. At the end of each feeding period, the mice were challenged with *Salmonella typhimurium*. The concentration of IgA against *Salmonella* in intestinal fluid was measured by ELISA on various days postchallenge. The unit was $OD_{493\ nm}$/100 min.

Remarks Orally fed *L. casei, L. acidophilus,* and yogurt enhanced the number of IgA-producing cells in the small intestine and the effect was most pronounced in the *L. casei* treatment. Significant increase in the concentration of secretory IgA that were specific for *S. typhimurium* in mice that were fed with *L. casei* and challenged with *S. typhimurium* was observed. Prolong feeding of *L. acidophilus* for up to 7 days resulted in a decrease in the number of IgA-producing cells. This was partly due to a decrease in the total number of lymphoid cells in the lamina propria of the intestine. Alterations of the epithelium in mice fed with *L. acidophilus* was observed but not in mice fed with *L. casei* and yogurt for up to 7 days.

The following two tables show effect of *B. breve* strain Yakult on virus-induced diarrhea and antibody production (12).

Passive Protection of Rotavirus-Induced Diarrhea in Mice

Group[a]	Immunization with SA-11[a]	Feeding with *B. breve*[b]	No. with Diarrhea/No. of pups (%)[c]
A	–	–	28/28 (100)
B	–	+	20/22 (90.9)
C	+	–	10/46 (21.7)[d]
D	+	+	1/39 (2.6)[d,e,f]

[a]Dams were not immunized (groups A, B) or immunized orally with single 10^6 plaque forming unit (pfu) dose of rotavirus SA-11 9–12 days before delivery (groups C, D).
[b]Dams were fed standard diet (groups A, C) or standard diet containing heat-killed *B. breve* strain Yakult at a concentration of 0.05% for 12 weeks before and 14 days after delivery (groups B, D).

[c]Five days after birth, pups in each litter were orally challenged with 2×10^6 pfu SA-11 and observed for diarrhea 1–4 days after inoculation.
[d]$p < 0.001$ vs. group A..
[e]$p < 0.001$ vs. group B..
[d]$p < 0.0081$ vs. group C..

Remarks Mouse pups born to and nursed by dams fed *B. breve* strain Yakult and immunized orally with rotavirus were strongly protected against rotavirus-induced diarrhea than those born to and nursed by dams immunized with rotavirus only.

IgA Antibody Titers in Milk and Feces from Dams Immunized Orally with SA-1 and Fed *B. breve* Strain Yakult (12)

Group	Immunization with SA-11[a]	Feeding with *B. breve*[b]	Anti-SA-11 IgA Titer in[c] Milk (U/ml)	Anti-SA-11 IgA Titer in[c] Feces (U/g)
A	–	–	13 ± 7	1.9 ± 1.1
B	–	+	13 ± 3	2.1 ± 0.4
C	+	–	51 ± 14	2.7 ± 1.3
D	+	+	85 ± 39[d]	8.2 ± 5.1[e]

[a]Dams were not immunized (groups A, B) or were immunized orally with single 10^6 pfu dose of SA-11 9–12 days before delivery (groups C, D).
[b]Dams were fed *B. breve* strain Yakult for 12 weeks before and 14 days after delivery.
[c]Five days after birth, pups in each litter were orally challenged with 2×10^6 pfu of SA-11, and IgA antibody titers in milk received by pups and feces were measured by ELISA 14 days after birth.
[d]$p < 0.02$ vs. group C.
[e]$p < 0.05$ vs. group C.

Remarks Both the levels of antirotavirus IgA in milk and those in feces of dams fed *B. breve* strain Yakult and immunized orally with rotavirus were significantly higher than those of dams immunized with rotavirus only, suggesting that oral administration of *B. breve* strain Yakult enhanced antigen-specific IgA antibody in the mammary gland and the intestine.

5.3.3 Effect on Phagocytes

Effect on Activity of Macrophages and Reticuloendothelial System (2)

Macrophage Function	Animal	Treatment	Days	Activity
Phagocytosis[a]	ICR mice	LC 9018[b]	0	270.8 ± 17.5 SRBC/100 cells
			2[c]	425.0 ± 75.0 SRBC/100 cells
Lysosomal enzyme[d]	ICR mice	LC 9018[b]	0	86.9 ± 16.5 munits/10^7 cells
			2[c]	165.3 ± 20.6 munits/10^7 cells
Carbon clearance[e]	ICR mice	LC 9018[b]	0	12.6 ± 1.5 min
			7[c]	3.1 ± 0.4 min
Cytostatic activity[f]	C57BL/6 mice	LC 9018[g]	1	70.0 ± 8.5 %[h]
			5[c]	37.9 ± 4.0 %[i]

[a]Assay for phagocytic activity: (a) Sheep red blood cells were washed twice with PBS and resuspended in PBS to a concentration of 1×10^9 cells/ml. Equal volume of 1:5 diluted anti-SRBC IgG was added and

incubated for 30 min at 37°C. (b) Peritoneal exudate cells (PEC) were obtained by washing out peritoneal cavities with Hanks' solution and then resuspended in Eagle MEM containing 10% fetal calf serum. The PEC suspension was placed on cover slip, incubated at 37°C for 90 min in 5% CO_2 in air. The cover slip was then washed to remove nonadherent cells. (c) The macrophages on the cover slip were incubated with anti-SRBC IgG coated SRBC at 37°C for 60 min. After treatment with 0.83% ammonium chloride to lyse the nonphagocytozed red cells, the number of SRBC phagocytozed by peritoneal macrophages was counted microscopically. Phagocytic activity was shown as the number of SRBC phagocytozed per 100 macrophages.

[b]Heat-killed *L.casei* strain Shirota (LC 9018) was injected intraperitoneally at 1 mg/mouse.

[c]The day when the maximum activity was measured. The activity of untreated mice (day 0) served as the control.

[d]Assay for lysosomal enzyme activity: (a) Macrophages (made up 90% of PEC harvested) were suspended in 100 mM phosphate buffer (pH 6) containing 2% Triton X-100 and subjected to four freezings and thawings. The solution was centrifuged for 20 min at 10,000 rpm, and the supernatant was used as enzyme solution. (b) The activity of acid phosphatase, one of the lysosomal enzymes, was measured using *p*-nitrophenyl phosphate as substrate. The reaction mixture contained 0.8 ml of 100 mM citrate buffer (pH 5.6), 0.1 ml of 50 mM substrate, and 0.1 ml enzyme solution. After incubation for 15 min at 37°C, 5 ml of 0.05 N sodium chloride was added, and the optical density at 405 nm was measured. One unit was expressed as the liberation of 1 μmol of *p*-nitrophenol per min at 37°C.

[e]Assay for carbon clearance rate: Carbon was injected into tail vein of the animal. At time intervals, 20 μl of blood was taken by capillary and added into 4 ml of 0.1% sodium carbonate. The carbon concentration was determined by optical density at 675 nm. The clearance rate of carbon ($t_{½}$) was calculated.

$$t_{½} = (t_2 - t_1)(½\mathrm{OD}t_1)/(\mathrm{OD}t_1 - \mathrm{OD}t_2)$$

where $\mathrm{OD}t_1$ and $\mathrm{OD}t_2$ is carbon concentration measured at time t_1 and t_2, respectively.

[f]Assay for cytostatic activity of PEC: Mixture of 100 μl of 1×10^4/ml EL4 leukemic cells and 100 μl of PEC at an appropriate concentration in RPMLI-1640 medium supplemented with 10% fetal calf serum (FCS) and 10 mm HEPES was seeded in tissue culture plate (Microtest II, Falcon 3042). The cells were cultured for 64 h in 5% CO_2 in air at 37°C. Then 1 μCi of ^3H-thymidine was added and incubated for another 8 h. Cells were collected on a glass filter and the radioactivity (cpm) incorporated into the tumor cells was determined by a liquid scintillation counter. The cytostatic activity of PEC was the inverse function of the percent incorporation of ^3H into the tumor cells, which was calculated from:

$$(\mathrm{cpm}_{(EL4 + PEC)} - \mathrm{cpm}_{(PEC)})100\% / \mathrm{cpm}_{(EL4)}$$

[g]Heat-killed *L. casei* strain Shirota (LC 9018) was injected intraperitoneally at 250 μg/mouse.

[h]Peritoneal exudate contained 47.3% macrophages, 1.9% lymphocytes, and 50.8% polymophornuclear cells.

[i]Peritoneal exudate contained 82.8% macrophages, 4.9% lymphocytes, and 12.3% polymophornuclear cells.

Remarks By intraperitoneal injection of heat-killed *L. casei* strain Shirota (LC 9018), the phagocytic activity of macrophages on sheep red blood cells, and acid phosphatase activity (one of the lysosomal enzymes) increased significantly. The phagocytic function of reticuloendotherial system was markedly stimulated. LC 9018 stimulated the cytostatic activity of peritoneal exudate cells (PEC) on EL4 leukemic cells. The enhancement of cytostatic activity was in parallel with the increase of macrophages in PEC, suggesting that the cytostatic activity of PEC depended on activated macrophages induced by LC 9018.

Effect on Activity of Peritoneal Macrophage and Reticuloendothelial System (7)

Macrophage Function	Feeding Treatment[a]	Day[b]	Activity
β-Galactosidase[c]	Nonfermented milk	3	19.90 ± 2.1
		5	19.90 ± 2.1
	L. casei fermented milk	3	58.10 ± 8.2
	L. acidophilus fermented milk	3	19.90 ± 2.1
	L. casei + *L. acidophilus* ferm. milk	5	49.40 ± 1.7
Phagocytosis[d]	Nonfermented milk	2	21 ± 3
		3	21 ± 3
	L. casei fermented milk	3	65.5 ± 5
	L. acidophilus fermented milk	3	64.1 ± 5
	L. casei + *L. acidophilus* ferm. milk	2	74.4 ± 6
Carbon clearance[e]	Nonfermented milk	5	9.90 ± 0.5
	Nonfermented milk supernatant[f]	5	10.00 ± 0.5
	L. casei fermented milk	5	0.27 ± 0.5
	L. casei fermented milk supernatant[f]	5	7.91 ± 1.2
	L. acidophilus fermented milk	5	0.11 ± 0.2
	L. acidophilus ferm. milk supernatant	5	1.82 ± 0.06
	L. casei + *L. acidophilus* ferm. milk	5	0.23 ± 0.3
	L. casei + *L. acidophilus* milk supern.	5	1.83 ± 0.03

[a]Swiss albino mice, 20–30 g (from the Centro de Referencia para Lactobacilos, Chacabuco, Argentina). Each data point is the average of 8–10 mice. The tested animals were fed with milk or fermented milk containing 100 μg cell protein/day (2.4×10^9 cells/day).

[b]Day when maximum activity was measured.

[c]The activity of enzyme released from peritoneal macrophages was expressed as nmols *o*-NPG liberated from the substrates *p*-nitrophenyl-β-D-glucurinide and *o*-nitrophenyl-β-D-galactopyranoside (Sigma, St. Louis, MO) per hour by 10^6 cells.

[d]Peritoneal macrophages (1×10^6 cells/ml) isolated from mice were incubated with Lactobacilli and *Salmonella typhi* (1×10^7 cells/ml) at 37°C for 5 min. The phagocytozing bacteria were counted microscopically after incubation and the phagocytic activity was expressed as percent phagocytosis.

[e]A solution of 8 mg/100 ml colloidal carbon was injected into the tail vein of mice; 50 μl blood was extracted by capillary from retroorbital venous plexus at intervals and added to 2 ml 0.1% $NaCO_3$. Carbon concentration was determined by optical density at 675 nm. Carbon clearance rate ($t_{½}$) was calculated as

$$t_{½} = (t_2 - t_1)(½\mathrm{OD}_{t_1})/(\mathrm{OD}_{t_1} - \mathrm{OD}_{t_2})$$

where $\mathrm{OD}t_1$ and $\mathrm{OD}t_2$ is carbon concentration measured at time t_1 and t_2, respectively.

[f]Supernatant of milk and fermented milk, centrifuged at 10,000*g* for 10 min.

Remarks Orally administered *L. casei* and/or *L. acidophilus* resulted in an increase in both the *in vitro* phagocytic activity of peritoneal macrophages and in the carbon clearance activity in Swiss albino mice. Such enhancement of the immune response might be partly due to substances produced by these organisms during the fermentation process, as reflected in the results obtained with the supernatant of the fermented milk culture.

Effect on Phagocytic Activity of Phagocytes and Fecal Counts of Probiotic Bacteria (9)

Treatment[a]	Granulocyte[b] (% phagocytic activity)	Monocyte[b] (% phagocytic activity)	Latobacilli Count[c] ($\log_{10}$ CFU/g feces)	Bifidobacteria Count[c] ($\log_{10}$ CFU/g feces)
Lactobacilli				
Before[d]	46.56 ± 2.75	37.95 ± 3.85	5.24 ± 0.39	8.29 ± 0.26
During[e] (3 wk)	84.98 ± 3.58	50.60 ± 5.50	6.46 ± 0.26[f]	8.41 ± 0.20
After[g] (12 d)	—	—	4.98 ± 0.38	8.13 ± 0.35
(6 wk)	73.98 ± 3.03	44.00 ± 4.95	—	—
Bifidobacteria				
Before[d]	41.25 ± 3.03	32.73 ± 4.68	5.52 ± 0.49	7.82 ± 0.56
During[e] (3 wk)	86.08 ± 4.95	48.13 ± 4.40	5.53 ± 0.41	9.00 ± 0.07[f]
After[g] (12 d)	—	—	5.74 ± 0.50	8.21 ± 0.32
(6 wk)	60.50 ± 5.50	39.88 ± 2.75	—	—

[a]Healthy adult human volunteers (12 females, 16 males) from 23 to 62 years of age (mean 36.9) were randomly divided into two groups. They consumed 120 ml of milk or fermented milk three times daily.

[b]Blood samples were collected in EDTA–blood collection tubes. Phagocyte activity was determined using flow cytometry and fluorescein isothiocyanate-labeled opsonized *E. coli* (PHAGOTEST, Becton-Dickinson, Basel, Switzerland): 100 μl fresh heparinized blood were mixed with 20 μl *E. coli* suspension (1×10^9/ml) such that the ratio of bacteria to leukocytes was 20:1 (v/v). The mixture was incubated for 10 min at 37°C with shaking. After quenching to remove free or attached bacterial cells, the leukocytes together with ingested bacteria were lysed and fixed, and DNA stained with propidium iodide. Measurements were with a FACScan flow cytometer using a blue-green excitation light (488 nm). During data acquisition with LYSIS II Software (Becton-Dickinson), a live gate was set in the red fluorescence histogram such that only those positive for propidium iodide staining were considered, thus corresponding to human diploid cells.

[c]Freshly passed fecal samples were analyzed. The lactobacilli were counted on MRS agar (Difco, Detroit, MI) with antibiotics (0.8 mg/ml of phosphomycine, 0.93 mg/ml of sulfanethoxazole, and 50 μg/ml of trimethoprim). Bifidobacteria were counted on Eugon agar (Becton-Dickinson) with tomato juice. Plates were incubated anaerobically for 48 h at 37°C. The lactobacilli were identified by cell morphology, and carbohydrate profile was determined by API 50CHL (Bio Mérieux, Marcy L'Etoile, France). Bifidobacteria were counted after identification by cell morphology and API 32A testing (Bio Mérieux).

[d]Received milk.

[e]Received a fermented milk supplemented with 7×10^{10} CFU/day *L. acidophilus* strain La 1 or 1×10^{10} CFU/day *B. bifidum* strain Bb12.

[f]$p < 0.05$ using a covariance analysis (ANCOVA).

[g]Received milk.

Remarks No modifications of lymphocyte subpopulations were detected (data not shown, for T-lymphocytes, B-lymphocytes, helper-inducer, suppressor-cytotoxic lymphocytes, and natural killer cells). Phagocytosis of *E. coli* in vitro was enhanced after the administration of both fermented products. The increment in phagocytosis was coincident with fecal colonization by the bacteria. Thus, nonspecific, anti-infective mechanisms of defense can be enhanced by ingestion of specific lactic acid bacteria strains.

5.3.4 Effect on T-lymphocyte

Effect on Cytotoxic Activity of Natural Killer (NK) Cells and Intraperitoneal Cytotoxic T-Lymphocytes (CTL) (8)

			NK activity (% lysis)[d]				
			Time after Oral Administration of PG				
Treatment[a]	Target Cells[b]	Effector Cells[c]	24 h	1 wk	2 wk	3 wk	Control
Single dose[e]	K562	Spleen	16.5 ± 20.4	22.5 ± 11.7	13.4 ± 8.7	7.1 ± 8.1	10.0 ± 11.4
		MLN	13.3 ± 18.2	16.4 ± 10.9	10.9 ± 4.6	13.8 ± 12.0	12.0 ± 11.4
	Mitsukaido	Spleen	11.2 ± 6.9	16.5 ± 9.7	7.5 ± 5.9	7.6 ± 5.4	11.1 ± 11.0
		MLN	9.9 ± 6.8	10.2 ± 10.0	6.0 ± 4.8	14.4 ± 16.0	7.3 ± 5 .3
Fed daily[f]	K562	Spleen				22.5 ± 18.6	24.6 ± 9.9
		MLN				30.7 ± 4.0[g]	9.6 ± 15.4
	Mitsukaido	Spleen				25.7 ± 4.7[g]	5.0 ± 9.2
		MLN				33.2 ± 11.5[h]	7.7 ± 9.3
			Cytotoxic T-lymphocyte activity (%)[i]				
Fed daily[f]	P815	Peritoneal	9.1[h]				7.1

[a]C57BL/6CrSlc male, 6–9 weeks old, specific pathogen-free mice (Japan SLC, Shizuoka, Japan) were used.
[b]K562 cell line was derived from chronic leukemia in humans; Mitsukaido cells from B lymphoid oncocyte of pigs; (P815 cells were mouse mastocytoma cells). The ratio of effector cells to target cells was 40:1, except in the test for cytotoxic activity, where the ratio was 50:1.
[c]Lymphocytes were derived from spleen and mesenteric lymph node (MLN). The tissues were mashed on 200 mesh, and cells were suspended in RPMI 1640 medium. Cells were carefully layered onto Histopaque-1077 (Sigma, USA) and centrifuged at 400*g* for 30 min at room temperature. The layer of lymphocytes was transferred into clean tube, washed twice with RPMI 1640 medium supplemented with 10% fetal calf serum, and resuspended in the same medium at a concentration of 2×10^6 cells/ml. The lymphocytes sensitized with P815 cells *in vivo* were harvested from peritoneal cavities in 10 ml RPMI 1640 medium containing 1 unit/ml sodium heparin with an injection pump. The cells were washed with RPMI 1640 medium, transferred to a plastic petri dish, and incubated at 37°C for 2 h under 5% CO_2 in air. Only the plastic adherent cells were used for the effectors of the CTL activity test.
[d]Assay for cytotoxicity of NK cells: (a) The target cells were washed once with RPMI 1640 culture medium and suspended in 1 ml of the same medium; 10 μl of 2,7-abis (2-carboxy-ethyl)-5,6-carboxyfluorescein, acetaxymethylester (BCECF-AM) was added to the cell suspension and incubated at 37°C for 1 h under 5% CO_2. (b) After that cells were washed 3 times with RPMI 1640, adjusted to 1×10 cells per ml in the same medium supplemented with 10% FCS. (c) The target cells were co-cultured with the effector cells at a ratio of 40:1 in a 96-well plate, and incubated at 37°C in 5% CO_2 in air. (d) After 4 h, latex beads up to 2% of the medium were added. The medium was stirred lightly, transferred to an Epicon plate (Idexx Ltd., Portland, ME), and fluorescence was determined by fluorescence concentration analyzer (Idexx). (e) Tridon × 100 was added to the target cells at 0.5% concentration and used as positive control. The target cells in RPMI 1640 medium supplemented with 10% FCS was used as a negative control. (f) The rate of activity of NK lymphocytes was determined:

$$\text{Percent lysis} = \frac{\text{(negative control value)} - \text{(tested value)}}{\text{(negative control value)} - \text{(positive control value)}} \times 100$$

[e]Peptidoglycan (PG) was prepared as follow: The P2-91 strain of *B. thermophilum* was cultured for 16 h at 37°C in Brigges liver broth under anaerobic conditions. The bacteria were then disrupted by a French press and washed twice with Sørensen's phosphate buffer (pH 6.2) by centrifugation at 5000 rpm for 30 min and precipitate was suspended in the same buffer. The suspension was treated with 0.01% lysozyme and 0.05% pronase for 48 h at 37°C. One milliliter of PG preparation contained the cell wall digest from 10^{12} native *B. thermophilum* cells. The PG preparation was stirred for 30 min and sonicated for 15 min. Then the solution was adjusted to 500

μg (content of hexosamine 2.9%)/0.5 ml PBS and was administered directly to the stomach with a stainless steel canula. Sterile PBS was used as the control.
[f]A 50-μg freeze-dried PG preparation was mixed with 5 g commercial powder feed (Oriental Yeast, Japan) and fed to mice daily for 3 weeks. The control group was fed only with commercial feed.
[g]Significantly different, $p < 0.05$.
[h]Significantly different, $p < 0.01$.
[i]The cytotoxic activity of CTL was determined as for NK activity (refer to footnote *d*) except that P815 cells were used as the target cells, and the effector cells to target cells ratio was 50:1.

Remarks The NK cells from the spleen and mesenteric lymph node of mice that were continuously fed with PG for 3 weeks showed a significantly higher rate of cytolysis than those from the control group. However, a single oral administration of PG had no significant effect on NK activity. The activity of peritoneally sensitized CTL of mice that were continuously fed with PG was significantly higher that the control group. These results indicated that orally administered PG from *B. thermophilum* enhanced the cytotoxic activity of mice.

Effect on Cytotoxic Activity of NK Cells (3)

Mouse[a]	Treatment	Cytolytic Activity (%)[b]	Tumor Wt (g)
C3H/He	Control[c]	19.4 ± 5.2	
	LC 9018[c]	53.3 ± 1.6	
CBA/N	Control[c]	27.4 ± 5.7	
	LC 9018[e]	49.7 ± 2.2	
C57BL/6	Control[e]	9.6 ± 2.9	
	LC 9018[e]	40.6 ± 5.4	
BALB/c	Control[c]	8.7 ± 1.8	
	LC 9018[c]	30.2 ± 3.1	
BALB/c	Control[d]	4.3 ± 0.7	
	LC 9018[d]	34.3 ± 1.5	
BALB/c-MA 10-day[e]	Control[d]	9.7 ± 1.3	0.19 ± 0.03
	LC 9018[d]	19.4 ± 2.6	0.11 ± 0.02
BALB/c-MA 20-day[e]	Control[d]	9.4 ± 0.8	1.05 ± 0.17
	LC 9018[d]	19.6 ± 1.4	0.62 ± 0.2
BALB/c	Control[d] + complement[f]	7.5 ± 1.6	
	Cont + compl + anti-Thy Ab[g]	6.7 ± 0.5	
	Cont + compl + anti-asialo Ab[h]	–0.2 ± 0.3	
	LC 9018[d] + complement	24.5 ± 0.9	
	LC + compl + anti-Thy Ab	13.6 ± 0.4	
	LC + compl + anti-asialo Ab	–0.4 ± 0.08	

[a]Male BALB/c, C57BL/6, and C3H/He mice from Shizuoka Agricultural Cooperative Association for Experimental Animals, Shizuoka. Male CBA/N mice from CLEA Japan Inc., Tokyo.
[b]Assay for cytotoxicity of NK cells: (a) Tumor cells YAC-1 derived from Moloney virus-induced lymphoma in A/Sn mouse (5×10^6) suspended in Hanks' solution was labeled for 60 min at 37°C with 100 μCi of $Na_2\,^{51}CrO_4$ (New England Nuclear, Boston, Mass.). The cells were washed three times and adjusted to 2×10^5 cells/ml in RPMI medium containing 10% FCS and 10 mM HEPES. 0.1 ml of YAC cells was added to round-bottom tissue culture plate (76-013-05, Linbro Scientific, Hamden, CT). (b) Appropriate concentration of spleen cells in 0.1 ml volume was added. The plate was centrifuged for 5

min at 500 rpm, incubated for 4 h at 37°C in 5% CO_2 in air. The cell mixture was further centrifuged for 7 min at 1500 rpm, and 100 μl of the supernatant was harvested. (c) Radioactivity was counted in a gamma counter (400 CGD, Packard Instrument Co., Downers Grove, IL). The cytotoxic activity was expressed as percentage specific ^{51}Cr released:

$$\text{Specific lysis (\%)} = \frac{(\text{experimental } ^{51}\text{Cr released}) - (\text{spontaneous } ^{51}\text{Cr released}) \times 100}{(\text{total } ^{51}\text{Cr}) - (\text{spontaneous } ^{51}\text{Cr released})}$$

The spontaneous ^{51}Cr released from target cells incubated in medium alone was less than 10% of total ^{51}Cr.

[c]A 0.1-ml saline (control) solution or 100 μg LC 9018 per day for 3 days, intravenous. LC 9018 was heat-killed *L. casei* obtained from Yakult, Tokyo, Japan.

[d]A 250-μl saline (control) solution or 250 μg LC 9018 was injected intravenously 3 days before NK assay.

[e]Meth A cells (1 × 10^5 cells/mouse) originated in a BALB/c mouse were inoculated subcutanously 10 or 20 days before NK assay.

[f]1:1 volume 1:10 diluted guinea pig complement.

[g]One milliliter of anti-Thy 1.2 monoclonal IgM antibody (F7D5) (from Olac Ltd, Oxon, England) was incubated with murine spleen cells (2 × 10^7) for 40 min at 37°C before complement was added.

[h]One milliliter of rabbit anti-asialo GM1 antibody (Wako Pure Chemical Industries Ltd., Osaka, Japan) was incubated with murine spleen cells (2 × 10^7) for 40 min at 37°C before complement was added.

Remarks Heat-killed *L. casei* strain Shirota (LC 9018) augmented the NK cell activity of spleen from BALB/c, C3H/He, CBA/N, and C57BL/6 mice injected intravenously with the bacterium. The cytolytic activity of spleen cells after the injection of 250 μg LC 9018/mouse peaked on day 3. The increased NK activity induced by LC 9018 was observed in normal as well as tumor (Meth A)-bearing mice. The tumor growth was also suppressed by LC 9018 treatment.

In vitro treatment with anti-asialo GM1 antibody plus complement completely abrogated the LC 9018 augmented NK cell activity. The treatment with anti-thy 1.2 monoclonal antibody plus complement reduced the NK activity to half. This suggests that all NK cells in the spleen derived from LC-9018-treated mice carry the surface antigen, glycosphingolipid asialo GM1. The NK cell population may consist of two subpopulation; one being Thy 1-positive whereas the other is Thy 1 negative.

5.3.5 Effect on Hematopoiesis

Effect on Hematopoietic Response (13)

Days[a]	Spleen Macrophage (× 10^7/spleen)	Spleen Lymphocyte (× 10^7/spleen)	Spleen[b] PMN Cells (× 10^7/spleen)	Spleen[c] CFUm (× 10^3/spleen)	Femur[c] CFUm (× 10^3/spleen)	Serum[d] CSA (CFU/10^5 BMC)
0	4.79	15.32	0.40	0.34 ± 0.14	7.64 ± 1.22	1.31
1	7.10	5.86	0.04	0.50	17.27 ± 0.99	1.31
3	8.39	11.46	0.89	4.54 ± 0.43	35.26 ± 1.27	1.31
5	14.07	10.88	0.40	6.80 ± 0.51	9.94 ± 0.72	1.31
7	16.74	5.77	0.88	4.49 ± 0.18	9.04 ± 1.40	—
9	—	—	—	—	—	4.50 ± 0.75
10	10.74	8.08	0.67	3.44 ± 0.25	2.21 ± 0.95	—

(continued)

12	—	—	—	—	—	5.63 ± 1.38
14	5.28	12.43	0.40	—	—	—
18	—	—	—	—	—	26.75 ± 1.94
24	—	—	—	—	—	4.44 ± 1.19

[a]Days after subcutaneous injection of 0.5 mg lyophilized heat-killed *L. casei* strain Shirota (provided by Yakult Central Institute for Microbiology Research, Tokyo) suspended in saline, into 6-week-old male BALB/c mice, obtained from Shizuoka Agricultural Cooperative for Experimental Animals, Hamamatsu, Japan.
[b]Polymorphonuclear cells in spleen.
[c]To determine macrophage colony-forming cells (CFUm), L-cell-conditioned medium (LCM) was prepared as follows: mouse L 929 cells were seeded in a 75-cm^2 culture flask at 2×10^5 cells in 40 ml Dulbecco modified Eagle medium supplemented with 10% horse serum (GIBCO, USA) and antibiotics, and incubated at 37°C in 5% CO_2 for 7 days. The cells were removed by centrifugation, and the supernatant was filter sterilized and stored at –20°C. The number of macrophage progenitor cells that would proliferate in response to LCM was determined by the two-layer agar culture method: 1.0 ml Dulbecco modified Eagle medium containing 20% LCM, 0.5% agar, and antibiotics were allowed to set in a petri dish (35 by 10 mm). Leukocytes from either spleens or femurs were suspended at 5×10^6 (spleen) or 10^5 (femur) cells per milliliter in Dulbecco modified Eagle medium containing 0.3% agar and antibiotics. A 1-ml leukocyte suspension was added on top of the 0.5% agar underlayer, and the plate was incubated at 37°C in 10% CO_2 in air. Colonies (more than 50 cells) were scored microscopically on day 7 of incubation and expressed as macrophage colony-forming units (CFUm).
[d]Assay for colony-stimulating activity (CSA) of serum; test sera from groups of three mice were pooled, filtered through 0.22-μm membrane filter, and stored at –20°C. Each sample was added to the underlayer of a two-layer semisolid agar culture system (refer to the method for the determination of macrophage colony-forming cells above), which did not contain LCM. Bone marrow cells (10^5 BMC) from untreated mice were added to the upper layer. The dish was incubated at 37°C in 10% CO_2 for 7 days. Colonies were scored and cells in the colonies were identified by Giemsa staining. Colony-stimulating activity was expressed as number of colonies/10^5 bone marrow cells.

Remarks After the administration of heat-killed *L. casei* strain Shirota (LC 9018), the ratio of monocyte–macrophage in the spleen increased, reaching its peak on day 5–7. The number of progenitor cells that formed macrophage colonies in the spleen increased after injection of LC 9018, showing a peak response on day 5. The number of macrophage colonies in the femur of mice pretreated with LC 9018 showed a temporary increase on day 3 after injection. Colony-stimulating activity was detected in sera of mice administered LC 9018, and the colonies produced were of three types: granulocyte (8%), macrophage (56%), and granulocyte–macrophage (36%).

5.3.6 Effects on Allergic Immune Responses in Mice (4)

Inhibition of Antigen-Induced IgE Secretion

	Amount of Dye (mg/ml)[a]
Control[b]	2451 ± 1192
0.05% LC-MF[c]	467 ± 390[d]
0.1% LC-MF[c]	820 ± 1139[e]

[a]BALB/c mice ($n = 5$) were intraperitoneally injected on day 0 and 14 with 20 μg of ovalbumin (OA, Sigma Chemical Co., St. Louis, MO) and 2 mg of $Al(OH)_3$, and sera were collected from the mice on day 21 after the injection of OA. Diluted sera (0.1 ml) from each mouse ($n = 6$) were intradermally injected on the shaved dorsal skin of Wistar rats, and, 24 h later, each rat was injected intravenously with 1 mg of OA in 1 ml of 0.5% Evans blue solution. After 30 min, the reaction was read from the reflected skin. The reaction was expressed as the amount of dye extracted from the skin spot.

[b]Standard diet.
[c]Diet containing 0.05 or 0.1% *L. casei* strain Shirota were given for 21 days after treatment of the mice with OA.
[d]$p < 0.05$.
[e]$p < 0.01$.

Remarks Ovalbumin-specific IgE production was inhibited dramatically by feeding mice with *L. casei*-containing diet after treatment of the mice with OA.

Effect on Cytokine Production by Spleen Cells *in vitro*[a]

Cytokine	Control	*L.casei*
IFN γ (pg/ml)	665 ± 71	986 ± 55[b]
IL-2 (pg/ml)	105 ± 18	159 ± 3[b]
IL-4 (pg/ml)	287 ± 4	149 ± 15[b]
IL-5 (pg/ml)	599 ± 40	298 ± 25[b]
IL-6 (pg/ml)	3757 ± 352	2453 ± 77[b]
IL-10 (pg/ml)	17 ± 5	3 ± 1[b]
IL-12 (pg/ml)	135 ± 50	880 ± 30[b]

[a]BALB/c mice ($n = 6$) were intraperitoneally injected on day 0 and 14 with 20 μg of OA and 2 mg of $Al(OH)_3$, at a total volume of 0.2 ml. The mice were then fed a diet containing (wt/wt) 0.05% *L. casei* strain Shirota for 21 days. Spleen cells were collected on day 21 and were cocultured with OA (final 100 μg/ml) for 3 days. The amounts of the cytokines in the supernatants were measured by ELISA. Data are expressed as the mean value (± SD) for triplicate cultures.
[b]$p < 0.01$.

Remarks In the mice fed *L. casei* strain Shirota, the production of Th1 cell-associated cytokines by the spleen cells, such as interferon-γ and interleukin-2, in response to OA was higher than that by the spleen cells from the control group. In contrast, the production of Th2 cell-associated cytokines, such as IL-4, 5, 6 and IL-10, by spleen cells in the group fed *L. casei* strain Shirota was lower than that by the cells from the control group. Furthermore, the interleukin-12 production by the spleen cells from the mice fed *L. casei* was also higher than that by the control group.

REFERENCES

1. Araki, Y., Nakatani, T., Nakayama, K., and Ito, E. Occurrence of N-nonsubstituted glucosamine residues in peptidoglycan of lysozyme-resistant cell walls from *Bacillus cereus*. *J. Biol. Chem.* **247:** 6312, 1972.
2. Kato, I., Yokokura, T., and Mutai, M. Macrophage activation by *Lactobacillus casei* in mice. *Microbiol. Immunol.* **27**(7): 611–618, 1983.
3. Kato, I., Yokokura, T., and Mutai, M. Augmentation of mouse natural killer cell activity by *Lactobacillus casei* and its surface antigens. *Microbiol. Immunol.* **28**(2): 209–217, 1984.
4. Matsuzaki, T., Yamazaki, R., Hashimoto, S., and Yokokura, T. The effect of oral feeding of *Lactobacillus casei* strain Shirota on immunoglobulin E production in mice. *J. Dairy Sci.* **81:** 48–53, 1998.

5. Nomoto, K., Miake, S., Yokokura, T., Shimizu, T., Mutai, M., and Nomoto, K. Systemic augmentation of host defense mechanism by subcutaneous administration of *Lactobacillus casei* in mice. *J. Clin. Lab. Immunol.* **18:** 43–48, 1985.
6. Perdigón, G., Alvarez, S., Rachid, M., Aguero, G., and Gobbato, N. Immune system stimulation by probiotics. *J. Dairy Sci.* **78:** 1597–1606, 1995.
7. Perdigón, G., Nader de Macias, M. E., Alvarez, S., Oliver, G., and de Ruiz Holgado, A.P. Systemic augmentation of the immune response in mice by feeding fermented milks with *Lactobacillus casei* and *Lactobacillus acidophilus. Immunology* **63:** 17–23, 1988.
8. Sasaki, T., Fukami, S., and Namioka, S. Enhancement of cytotoxic activity of lymphocyte in mice by oral administration of peptidoglycan (PG) derived from *Bifidobacterium thermophilum. J. Vet. Med. Sci.* **56**(6): 1129–1133, 1994.
9. Schiffrin, E. J., Rochat, F., Link-Amster, H., Aeschlimann, J. M., and Donnet-Hughes, A. Immunomodulation of human blood cells following the ingestion of lactic acid bacteria. *J. Dairy Sci.* **78:** 491–497, 1995.
10. Shimizu, T., Mifuchi, I., and Yokokura, T. Mitogenic effect of *Lactobacilli* on murine lymphocytes. *Chem. Pharm. Bull.* **29**(12): 3731–3734, 1981.
11. Yasui, H., and Ohwaki, M. Enhancement of immune response in Peyer's patch cells cultured with *Bifidobacterium breve. J. Dairy Sci.* **74:** 1187–1195, 1991.
12. Yasui, H., Kiyoshima, J., and Ushijima, H. Passive protection against Rotavirus-induced diarrhea of mouse pups born to and nursed by dams fed *Bifidobacterium breve* YIT 4064. *J. Infect. Dis.* **172:** 403–409, 1995.
13. Yokokura, T., Nomoto, K., Shimizu, T., and Nomoto, K. Enhancement of hematopoietic response of mice by subcutaneous administration of *Lactobacillus casei. Infect. Immun.* **52**(1): 156–160, 1986.

5.4 ALTERATION OF INTESTINAL BACTERIAL METABOLIC ACTIVITY

Target	Subject	Probiotics	Route	Viability	Effect	Ref.
Intestinal bacterial enzyme activity	Healthy human	*L. acidophilus* N2, NCFM	Oral	Viable	Reduce activity of fecal β-glucuronidase, nitroreductase, azoreductase	1
Intestinal bacterial enzyme activity	Healthy human	Fermented milk *L. acidophilus* A1 *B. bifidum* B1 Lactic culture MO (*S. lactis* and *S. cremoris*)	Oral	Viable	Suppress nitroreductase activity, promote β-glucosidase activity. No effect on fermentation capacity, fecal β-galactosidase, β-glucuronidase, and azoreductase activity	3
Intestinal bacterial enzyme	Healthy female adults	*L. rhamnosus* GG	Oral	Viable	Reduced fecal β-glucuronidase, nitro-reductase, glycocholic	2

(*continued*)

Target	Subject	Probiotics	Route	Viability	Effect	Ref.
activity and bacterial metabolism					acid hydrolase activities. β-glucosidase and urease activity not altered. Reduce urinary excretion of *p*-cresol.	

Effect of Milk and *Lactobacillus* Feeding on Human Intestinal Bacterial Enzyme Activity (1)

	Specific Activity (μg/min/mg protein)[b]		
Days on Protocol[a]	β-glucuronidase	Nitroreductase	Azoreductase
Baseline 1[c]			
0	1.74 ± 0.12	6.2 ± 0.61	4.6 ± 0.50
10	1.60 ± 0.10	5.4 ± 0.88	4.3 ± 0.59
Milk feeding[d]			
20	2.14 ± 0.20	4.3 ± 0.41	3.9 ± 0.46
30	2.15 ± 0.41	4.7 ± 0.49	4.7 ± 0.47
40	2.00 ± 0.19	4.0 ± 0.32	4.3 ± 0.38
50	2.10 ± 0.15	4.1 ± 0.47	4.1 ± 0.38
Lactobacillus feeding[e]			
90	1.30 ± 0.16	3.2 ± 0.41	3.1 ± 0.29
100	1.10 ± 0.29	1.6 ± 0.30	2.1 ± 0.15
110	1.12 ± 0.19[g]	1.2 ± 0.17[h]	1.2 ± 0.15[i]
Baseline[f]			
120	1.25 ± 0.16	2.2 ± 0.30	2.0 ± 0.15
130	1.72 ± 1.00	2.8 ± 0.36	3.1 ± 0.34
140	1.90 ± 0.27	4.2 ± 0.32	4.1 ± 0.34

[a]Duration of the study was approximately 140 days. Twenty-one healthy subjects were recruited for the study, 16 women (mean age 27.9 y) and 5 men (mean age 29.2 y). All but 2 were omnivores; with negative history of diarrhea, constipation, bowel disorder, and use of antibiotic over the previous 2 months.

[b]Subjects provided 15 fecal specimens spaced approximately every 10 days. All fecal specimens were assayed for the three bacterial enzyme activity within 2 h of evacuation of feces.

[c]Subjects provided three fecal specimens, which were assayed for β-glucuronidase, nitroreductase, and azoreductase. Baseline 1 provided data on the subjects prior to any attempt to manipulate the intestinal flora.

[d]During this period, subjects were asked to come to the laboratory each day and drink 2 glasses (500 ml) of low-fat milk. Three fecal specimens at 10-day intervals were again collected and fecal enzyme assays done.

[e]Subjects were returned to baseline protocol for 1-month period for enzyme levels to stabilize, on the assumption that the addition of milk to the diet might alter fecal enzyme activity. Subjects were then asked to drink 2 glasses of milk per day and the milk was supplemented with either 2×10^6/ml viable *L. acidophilus* N2 or 2×10^6/ml of *L. acidophilus* NCFM. Both strains were isolated from human feces. The concentration of viable lactobacilli, 10^6/ml, used in this study is similar to that achieved in commercial acidophilus milk and in some yogurt products, which contain viable *L. acidophilus.*

[f]Subjects continued their normal diet (without milk plus *Lactobacillus*) for the final month.

[g]Paired *t*-test with baseline value: $p < 0.001$.

[h]Paired *t*-test with baseline value: $p < 0.002$.

[i]Paired *t*-test with baseline value: $p < 0.02$.

Remarks The three bacterial enzymes assayed can catalyze procarcinogen conversion to proximal carcinogens, which have been implicated in the development of colon cancer. The study demonstrates that oral administration of viable *L. acidophilus* of human origin can cause a reduction of two to fourfold in the activities of the three fecal enzymes in healthy individuals. Drinking milk alone caused no significant alterations.

Effect of Consumption of Fermented Milk on Human Fecal Enzyme Activity (3)

	Period 1[a]	Period 2[b] (IU/g N)	Period 3[a]
Azoreductase[c]	0.96 ± 0.19	1.44 ± 0.33	0.72 ± 0.12
Nitroreductase[c]	0.87 ± 0.13	0.54 ± 0.11[d]	0.57 ± 0.08[d]
β-glucuronidase[c]	33 ± 5	41 ± 6	31 ± 5

[a]During these periods, the subjects ate their normal diet but were not allowed to consume fermented dairy products.
[b]During this period, at the end of their 3 meals, the subjects ingested 100 g of the fermented dairy product. The fermented dairy product was a commercial fermented milk (Ofilus, Yoplait, Paris) containing *L. acidophilus* strain A1 (1×10^7/g), *B. bifidum* strain B1 (1×10^8/g), and mesophilic lactic culture MO (*S. lactis* and *S. cremoris,* 1×10^8/g).
[c]During the morning before the last day of each period, each 3 weeks long, the subjects evacuated their feces into containers in the laboratory, and these were immediately maintained in anaerobic conditions and three different enzyme activities were assayed for within 2 h of evacuation. The subjects were nine healthy adults (six men and three women aged 30–42 y) and were identified as lactose absorbers.
[d]Significantly different from period 1, $p < 0.05$.

Remarks Azoreductase and β-glucuronidase activities did not change throughout the three periods. In contrast, nitroreductase activity dropped significantly in period 2 and remained at a low level during period 3. This study confirms that regular consumption of a fermented dairy product containing viable *L. acidophilus* induces a decrease in fecal nitroreductase activity. Nitroreductase can convert nitrated aromatic compounds into potentially harmful amines and has been implicated in colonic carcinogenesis. The decrease in nitroreductase activity persisted 3 weeks after cessation of the dairy product ingestion, suggesting prolonged modifications of the colonic flora.

Effect of Consumption of Fermented Milk on Fermentation Capacity of Colonic Flora (3)

	Period 1[a]	Period 2[a]	Period 3[a]
Area under delta breath-hydrogen profile (ppm/h)[b]			
Right colon microflora	247 ± 42	167 ± 80	251 ± 66
Left colon microflora	40 ± 2	48 ± 2	59 ± 4
Area under delta breath-methane profile (ppm/h)[b]			
Right colon microflora	0	0	0
Left colon microflora	69 ± 21	54 ± 35	63 ± 33

(continued)

	Period 1[a]	Period 2[a]	Period 3[a]
β-galactosidase (IU/g N)	152 ± 18	118 ± 32	103 ± 19
β-glucosidase (IU/g N)	42 ± 6	91 ± 12[c]	40 ± 6

[a]Each period lasted 3 weeks and the treatments were as mentioned in the previous table.
[b]Hydrogen and methane in end-expiratory gas were measured at 30-min intervals for 8 h after receipt of 10 g lactulose per os (right colon microflora) or in breath by enema (left colon microflora). Fecal β-galactosidase and β-glucosidase activities were assayed before the last day of each period.
[c]Significantly different from periods 1 and 3, $p < 0.01$.

Remarks Hydrogen and methane excretion after ingestion of 10 g lactulose and after a 10-g lactulose enema were not affected by consumption of the dairy product. β-galactosidase activity did not change, but β-glucosidase activity significantly increased in period 2 and returned to baseline levels in period 3. The conclusion is that chronic ingestion of a fermented dairy product containing *L. acidophilus* and *B. bifidum* of human origin leads to metabolic modification of the colonic flora.

Effect of Oral Supplementation with *L. rhamnosus* GG Yogurt and Fiber on Fecal Enzyme Activities and Bacterial Metabolites in Urine of Female Humans (2)

	Supplementation[a,b]			
	Baseline	2 wk	4 wk	Follow-up
ß-Glucuronidase, nmol/(min.mg fecal protein)				
L. GG (14)	8.0 ± 2.4	5.7 ± 2.3†	5.2 ± 2.4†	8.5 ± 3.4
L. GG + fiber (16)	9.3 ± 2.4	7.2 ± 2.4†	6.9 ± 1.9†	8.5 ± 1.6
Placebo + fiber (12)	7.2 ± 1.1	6.7 ± 1.5	6.6 ± 1.6	7.0 ± 2.2
β-Glucosidase, nmol/(min.mg fecal protein)				
L. GG (14)	11.6 ± 6.1	12.7 ± 5.0	13.5 ± 7.3	11.7 ± 6.1
L. GG + fiber (16)	11.8 ± 5.8	12.2 ± 4.8	13.3 ± 4.4	13.7 ± 5.5
Placebo + fiber (12)	10.4 ± 5.1	10.7 ± 5.3	11.4 ± 4.2	12.0 ± 3.7
Glycocholic acid hydrolase, nmol/(min.mg fecal protein)				
L. GG (15)	11.9 ± 6.5	7.3 ± 5.0*	8.1 ± 4.6*	13.8 ± 4.3
L. GG + fiber (16)	15.8 ± 6.3	10.1 ± 5.1†	8.6 ± 4.4†	13.8 ± 7.8
Placebo + fiber (11)	11.5 ± 7.6	14.7 ± 6.4‡	14.2 ± 7.1‡	13.6 ± 6.5
Nitroreductase, nmol/(h.mg fecal protein)				
L. GG (13)	32.5 ± 4.3	24.0 ± 6.7†	23.9 ± 7.8†	33.3 ± 9.8
L. GG + fiber (16)	31.3 ± 9.8	23.2 ± 6.2†	22.2 ± 7.5†	28.5 ± 8.5
Placebo + fiber (10)	31.8 ± 6.9	30.0 ± 6.2‡	31.3 ± 6.9‡	32.9 ± 10.5
Urease, nmol/(min.mg fecal protein)				
L. GG (14)	50.9 ± 32.6	44.6 ± 26.6	46.0 ± 25.0	42.2 ± 25.0
L. GG + fiber (16)	50.6 ± 23.8	41.0 ± 23.5	47.2 ± 26.4	50.1 ± 23.2

(*continued*)

	Supplementation[a,b]			
	Baseline	2 wk	4 wk	Follow-up
Placebo + fiber (12)	52.0 ± 22.4	35.7 ± 17.3	34.1 ± 28.3*	41.2 ± 22.3
Phenol, μmol/d				
L. GG (11)	107 ± 66	91 ± 48	74 ± 37	102 ± 69
L. GG + fiber (11)	114 ± 82	134 ± 54	112 ± 78	121 ± 72
Placebo + fiber (11)	117 ± 47	137 ± 96	112 ± 79	120 ± 57
p-Cresol, μmol/d				
L. GG (11)	488 ± 225	448 ± 250	413 ± 197*	430 ± 203
L. GG + fiber (11)	416 ± 221	337 ± 181	328 ± 239*	384 ± 265
Placebo + fiber (11)	430 ± 231	395 ± 270	399 ± 203	391 ± 232

[a]A total of 64 healthy female students (mean age 24 y, range 20–41 y) participated in the study. The subjects were randomly divided into three groups. The first group received *L.* GG yogurt (2 × 150 ml/d, containing 10^{11} colony-forming units/l of *L.* GG). The second group received *L.* GG yogurt and a rye fiber product (30 g/d, equivalent to 9 g fiber/d). The third group received placebo yogurt (pasteurized) and fiber. The study lasted 4 weeks, with a preceding 2-week baseline period and a 2-week follow-up period. Values are means ± SD, with *n* in parentheses.

[b]Values significantly different from the appropriate baseline value are indicated by * $p < 0.05$, † $p < 0.01$ (multivariate ANOVA). Differences between the placebo + fiber group and the groups receiving *L.* GG are indicated by ‡, $p < 0.05$ (one-way Student–Newman–Keuls test).

Remarks The data demonstrated that *L. rhamnosus* can modify the activities of fecal β-glucuronidase, nitroreductase, and glucocholic acid hydrolase but not β-glucosidase and urease. These enzyme activities returned to baseline levels during the follow-up period. The addition of fiber had no effect on the enzyme activities. Production of *p*-cresol (product of anaerobic flora *Bacteroides fragilis* of colon) was suppressed, but phenol (product of aerobes *E. coli* in ileum) was not affected during the *L.* GG supplementation period.

REFERENCES

1. Goldin, B. R., and Gorbach, S. L. The effect of milk and lactobacillus feeding on human intestinal bacterial enzyme activity. *Am. J. Clin. Nutr.* **39:** 756–761, 1984.
2. Ling, W. H., Korpela, R., Mykkänen, H., Salminen, S., and Hänninen, O. *Lactobacillus* strain GG supplementation decreases colonic hydrolytic and reductive enzyme activities in healthy female adults. *J. Nutr.* **124:** 18–23, 1994.
3. Marteau, P., Pochart, P., Flourié, B., Pellier, P., Santos, L., Desjeux, J.-F., and Rambaud, J.-C. Effect of chronic ingestion of a fermented dairy product containing *Lactobacillus acidophilus* and *Bifidobacterium bifidum* on metabolic activities of the colonic flora in humans. *Am. J. Clin. Nutr.* **52:** 685–688, 1990.

5.5 ALTERATION OF MICROECOLOGY IN HUMAN INTESTINE

Subject	Probiotic	Route of Administration	Viability	Remarks	Ref.
Healthy human	*L. acidophilus*	Oral	Viable	Number of lactobacilli increased, but *E. coli* decreased during administration.	4
Healthy human	*L. casei* strain Shirota	Oral	Viable	Number of lactobacilli and bifidobacteria increased, resulted in lower β-glucoronidase and β-glucosidase activity. No effect on any immune parameters.	6
Healthy human	*B. longum*	Oral	Viable	Number of bifidobacteria increased, but bacteroides and lecithinase-negative clostridia decreased during feeding, resulted in lower fecal pH, ammonia content, and β-glucuronidase activity.	1
Patients with gastro-intestinal disorders	*B. bifidum* strain Yakult and *B. breve* strain Yakult	Oral	Viable	Number of bifidobacteria increase, but coliform bacteria and other aerobes showed tendency to decline.	7
Patients with leukemia	*Bifidobacterium* sp. and *L. acidophilus*	Oral	Viable	Intestinal bacterial and yeast population was restored to that before antileukemic drug administration and urine indican and blood endotoxin were suppressed.	3

5.5.1 Effect of *Lactobacillus* on Healthy Humans

Effect of *L. acidophilus* Supplements on Oropharyngeal and Intestinal Microflora (4)

	Day during Administration[a]				Day after Administration		
	0	2	4	7	2	4	9
Oropharynx (log number/ml saliva)[b]							
Streptococci	6.0–8.1	←→[d]					
Micrococci and Staphylococci	4.7–7.5	←→					
Haemophilus and Branhamella	4.0–8.0	←→					

(continued)

Effect of *L. acidophilus* Supplements on Oropharyngeal and Intestinal Microflora (4) (*cont*)

	Day during Administration[a]				Day after Administration		
	0	2	4	7	2	4	9
Lactobacilli	2.7–5.4		◄———	———	———	———►	
Anaerobic cocci	5.4–8.0		◄———	———	———	———►	
Fusobacteria	3.0–6.0		◄———	———	———	———►	
Bacteroides	4.4–8.2		◄———	———	———	———►	
Intestine (log number/g feces)[c]							
E. coli	3.0–8.0	2.9–7.7	1.5–7.3	2.5–7.4	3.0–7.5	3.0–8.1	3.5–8.0
Lactobacilli	3.4–7.0	4.1–7.9	4.2–8.1	4.4–9.2	4.3–7.3	3.6–7.1	3.6–7.2
Bacillus and Corynebacteria	2.2–6.3		◄———	———	———	———►	
Staphylococci and Streptococci	2.7–5.0		◄———	———	———	———►	
Enterococci	2.1–6.9		◄———	———	———	———►	
Anaerobic cocci	2.0–6.9		◄———	———	———	———►	
Clostridia	2.0–6.1		◄———	———	———	———►	
Eubacteria and Bifidobacteria	5.8–8.4		◄———	———	———	———►	
Fusobacteria	2.0–4.0		◄———	———	———	———►	
Bacteroides	7.0–9.2		◄———	———	———	———►	

[a]Ten healthy volunteers (six females and four males, 26–46 years old) were given a fermented milk product (Arla Acidofilus, Arla, Stockholm) containing 5×10^8–2×10^9 CFU/ml of *L. acidophilus* in a dose of 250 ml twice a day for 7 days.

[b]Mixed saliva samples were collected by spitting into sterile glass tubes immediately before taking *Lactobacillus* supplement, and then on second, fourth, and seventh day during administration, and again, 2, 4, and 9 days after the administration period.

[c]Stool specimens were collected in sterile plastic containers, placed in ice chests. The saliva and feces samples were suspended in prereduced peptone-yeast extract medium, diluted, and inoculated on selective media. The bacteria were identified using morphological, serological, and biochemical tests and gas–liquid chromatography. *L. acidophilus* was reidentified biochemically using API 50 CHL test system for Lactobacilli (API, France).

[d]Arrows indicate small changes in bacterial numbers within the range shown, during the test periods.

Remarks

1. Only minor changes in the number of most of the microorganisms in the oropharynx and intestine were observed.
2. In the intestine a decrease in the numbers of *E. coli* during the period taking *L. acidophilus* was observed, while there was a significant increase in the number of lactobacilli. Lactobacilli returned to the same level as before the study after the *L. acidophilus* administration was stopped.
3. The study suggested that *L. acidophilus* preparation should be taken continuously in order to maintain high level of lactobacillus in the intestine.

Parameters Measured before Test Period (wk 2), after 2 and 4 Weeks during Test Period (wk 4, and wk 6), and at End of Follow-up Period (wk 8) (6)

	Control Group[a]				Treatment Group[a]			
Parameter	wk 2	wk 4	wk 6	wk 8	wk 2	wk 4	wk 6	wk 8
Microbial counts[b] ($\log_{10}$ CFU/g feces)								
Total anaerobes	9.6 ± 0.4	9.9 ± 0.3	9.9 ± 0.2	9.9 ± 0.3	9.4 ± 0.4	9.9 ± 0.3	9.7 ± 0.3	9.6 ± 0.3
Bacteroidaceae	9.4 ± 0.4	9.6 ± 0.4	9.2 ± 0.4	9.6 ± 0.4	9.2 ± 0.4	9.6 ± 0.5	8.9 ± 0.4	9.5 ± 0.5
Bifidobacterium	9.1 ± 0.3	9.1 ± 0.6	9.3 ± 0.4	9.3 ± 0.5	8.8 ± 0.5	9.2 ± 0.5[d]	9.2 ± 0.4	8.9 ± 0.6
Lactobacillus casei Shirota	3.3 ± 2.1	2.9 ± 1.8	4.1 ± 1.8	3.8 ± 1.9	2.0 ± 0.0	7.5 ± 0.5[d]	7.5 ± 0.6[d]	2.0 ± 0.0
Lactobacillus total	7.3 ± 0.8	7.1 ± 1.0	6.7 ± 1.2	7.2 ± 0.9	6.8 ± 1.5	7.6 ± 0.7	7.4 ± 0.7	6.9 ± 1.0
Enterococcus	6.2 ± 0.8	5.6 ± 1.2	5.7 ± 0.9	5.2 ± 1.3	5.5 ± 0.8	4.7 ± 1.1	4.3 ± 1.5	4.3 ± 1.4
Clostridium	4.6 ± 1.6	4.5 ± 1.0	3.6 ± 1.8	3.3 ± 2.5	5.2 ± 1.0	4.7 ± 1.0	3.3 ± 2.0	3.7 ± 2.2
Bacilllus	3.1 ± 1.1	3.1 ± 1.1	2.6 ± 0.3	2.8 ± 1.1	2.9 ± 1.1	3.6 ± 0.8	3.0 ± 0.5	3.5 ± 0.4
Staphylococcus total	4.2 ± 2.2	2.6 ± 2.0	1.6 ± 1.0	2.4 ± 1.3	4.0 ± 1.8	2.2 ± 0.9	2.0 ± 1.5	1.1 ± 0.9
Staphylococcus aureus	1.0 ± 0.2	1.2 ± 1.0	0.8 ± 0.1	1.0 ± 0.8	1.2 ± 1.2	0.9 ± 0.3	1.1 ± 1.0	0.9 ± 0.6
Enterobacteriaceae	6.6 ± 0.6	6.6 ± 0.9	6.3 ± 1.0	6.4 ± 1.2	6.5 ± 1.5	7.3 ± 0.8	6.6 ± 1.1	6.8 ± 0.9
Yeast	1.9 ± 0.9	2.2 ± 1.3	2.1 ± 1.0	1.6 ± 1.2	1.5 ± 0.8	1.8 ± 1.1	1.4 ± 1.2	1.2 ± 1.1
Bacterial enzyme activities[c]								
Urease (× 10^1 U/10^{10} CFU)	112 ± 43	48 ± 30	34 ± 10	28 ± 7	139 ± 60	32 ± 9	64 ± 14	65 ± 18
β-glucuronidase (×10^{-2} U/10^{10} CFU)	80 ± 20	45 ± 7	41 ± 6	55 ± 14	167 ± 35	44 ± 6[d]	72 ± 13	123 ± 24
β-glucosidase (×10^{-2} U/10^{10} CFU)	443 ± 117	271 ± 316	215 ± 30	257 ± 50	747 ± 147	230 ± 53[d]	328 ± 76	548 ± 122
Tryptophanase (U/10^{10} CFU)	105 ± 24	61 ± 13	52 ± 8	71 ± 16	155 ± 33	48 ± 7	89 ± 19	131 ± 23

[a]Twenty healthy male subjects aged 40–65 years, were randomly assigned a control group (10) and a treatment group (10). During the first two weeks (1–2 wk) and last 2 weeks (7–8 wk), all subjects received a strictly controlled diet without fermented products. The same controlled diet was given during the intermediate 4-wk test period, but the treatment group received 3 times daily 100 ml of fermented milk containing 10^9 CFU/ml *L. casei* strain Shirota, whereas same amount of unfermented milk was given to the control group.

[b]Anaerobic bacteria were counted on the following agars after incubated anaerobically in gas tied plastic bags (Merck) at 37°C for 48–72 h: Reinforced Clostridial Agar (Oxoid CM151) supplemented with 5 g/l glucose, 75 ml/l sterile horse blood, and 75 ml (0.4%)/l China blue (RCB) agar for total anaerobic bacteria; RCB agar containing 80 mg/l kanamycin and 1 mg/l vancomycin for Bacteroidaceae; Eugon agar (BBL) supplemented with 10 g/l maltose (Merck), 400 ml vegetable juice (Campbell V8), and 5 ml/l propionic acid, pH at 6.0 ± 0.2 for *Bifidobacterium.* Aerobic and facultative anaerobic microorganisms were counted on the following agars after incubated aerobically at 37 or 24°C: Rogosa agar (Oxoid) for *Lactobacillus*; LBS agar (Oxoid) containing 10 mg/l vancomycin for *L. casei* Shirota; perfringens agar base (Oxoid) with 2 vials/l perfringens SFP selective supplement (Oxoid) and 50 ml/l egg yolk emulsion for *Clostridium*; Baird-Parker agar (Oxoid) containing egg yolk–tellurite emulsion for *Staphylococcus*; Slanetz and Bartley medium (Oxoid) for *Enterococcus*; violet red bile glucose agar (Oxoid) for *Enterobacteriaceae*; RCB agar containing 2 ml (1%)/l tellurite for *Bacillus*; Oxytetracycline–glucose–yeast extract agar (Oxoid) with oxytetracycline GYE selective supplement for yeasts.

[c]β-glucuronidase activity was determined as follows. Substrate solution (2-nitrophenyl-β-D-glucopyranoside) was added to a homogenized suspension of feces in PBS pH 6.5 (fecal dilution approx. 1:100). After incubation (20 min, 37°C) the enzyme reaction was stopped by the addition of 0.01 N NaOH. After centrifugation (10 min, 3000g), the formed *o*-nitrophenol was measured at 415 nm. β-Glucuronidase activity: substrate solution (phe-

nolphthalein-β-glucuronide) was added to a homogenized suspension of feces in PBS pH 6.5 (fecal dilution approx. 1:400). After 15 min incubation at 37°C the enzyme reaction was stopped by the addition of 0.2 M glycerin solution (pH 10.4). After centrifugation (10 min, 3000*g*), the formed phenolphthalein was measured at 553 nm. The tryptophanase activity was measured in fecal samples diluted with PBS (0.05 M, pH 7.0). To a 1-ml diluted sample 2 ml cold acetone was added. The mixture was centrifuged and the supernatant was discarded. Then 1 ml PBS and 0.05 ml toluene was added. The samples were shaken (60 rpm) for 10 min. A pyridoxal/bovine serum albumin/PBS solution and substrate (tryptophan/PBS) was added to the samples. After 20 min of incubation (37°C) color reagent (P-dimethyl aminobenzaldehyde) was added. This mixture was incubated for 10 min at room temperature and centrifuged. The optical density at 540 nm was measured. For the determination of urease activity a test kit with a modified manufacturers protocol was used (urea/ammonia test kit; Boehringer Mannheim, Mannheim, Germany). Urea and a buffer solution (triethanolamine, pH 8.0) containing 2-oxoglutarate, glutamate dehydrogenase, and NADH were added to a centrifuged (10 min, 3000*g*) fecal suspension. The amount of NADH oxidation was measured during 10 min at room temperature at 340 nm.

[d]Statistically significant difference ($p < 0.05$) between groups corrected for initial differences.

Remarks In comparison to the control group, the consumption of *L. casei* Shirota-fermented milk resulted in an increase of the *Lactobacillus* count in the feces in which the administered *L. casei* Shirota was predominant at the level of 10^7 CFU/g wet feces. This was associated with a significant increase in *Bifidobacterium* counts ($p < 0.05$). The β-glucuronidase and β-glucosidase activity per 10^{10} bacteria decreased significantly ($p < 0.05$) at the second of the 4-week test period with the consumption of *L. casei* Shirota-fermented milk. No treatment effects were observed for any of the immune parameters measured, including NK cell activity, phagocytosis, and cytokine production (data not shown). The results of this study suggest that consumption of *L. casei* Shirota-fermented milk is able to modulate the composition and metabolic activity of the intestinal flora and indicate that *L. casei* Shirota-fermented milk does not influence the immune system of healthy immunocompetent males.

5.5.2 Effect of *Bifidobacterium* on Healthy Humans (1)

Relative Percentage of Each Bacterial Group to Total Bacteria in *B. longum* Feeding Study

Fecal Microflora[c]	Before Feedings[a] (week)		During Feedings[b] (week)					After Feedings (weeks)	
	2nd	1st	1st	2nd	3rd	4th	5th	1st	2nd
Total bacteria[d]	6.9×10^{10}	7.9×10^{10}	7.7×10^{10}	9.6×10^{10}	5.7×10^{10}	6.2×10^{10}	6.2×10^{10}	8.2×10^{10}	7.3×10^{10}
Bacteroides	58.0	63.3	51.9	52.0	53.1	52.6	48.3	61.0	54.6
Eubacteria	20.3	7.6	13.0	14.6	21.2	21.0	19.3	12.2	19.1
Peptostreptococci	4.3	12.6	6.5	12.5	4.2	4.3	9.7	9.8	8.2
Bifidobacteria	14.5	10.2	28.2	14.6	21.3	21.1	22.6	17.1	16.4
Veillonellae	0.003	0.005	0.002	0.002	0.03	0.03	0.04	0.005	0.02
Lecithinase-positive clostridia	<0.0001	<0.001	0.02	0.001	<0.001	<0.001	<0.0001	<0.0001	<0.0001
Lecithinase-negative clostridia	2.9	6.3	10.4	6.2	1.0	0.3	0.02	0.02	1.6

(*continued*)

Fecal Microflora[c]	Before Feedings[a] (week)		During Feedings[b] (week)					After Feedings (weeks)	
	2nd	1st	1st	2nd	3rd	4th	5th	1st	2nd
Lactobacillus	0.007	0.001	0.001	0.06	0.001	0.001	0.0001	0.0001	0.002
Enterobacteria	0.009	0.03	0.01	0.04	0.03	0.03	0.04	0.02	0.14
Streptococci	0.003	<0.001	0.01	0.06	0.02	0.02	0.02	0.002	0.02

[a]Five healthy male adults aged 26–38 years old.
[b]Subjects were served 1 g *B. longum* preparation (obtained from Nikken Chemicals Co. Ltd., Tokyo) after each usual meal for 5 weeks. One gram of the bacterial preparation contained 2–5 × 10^9 viable *B. longum* strains F6-1-ES and 69-2bs, originally isolated from feces of a human breast-fed baby.
[c]Freshly voided feces were collected at the end of each week of the treatment. The samples were serial diluted from 10^{-1} to 10^{-8}. Aliquot from appropriate dilutions was spread on nonselective and selective agar plates. Then, the dilutions were heated at 80°C for 10 min and inoculated on EG, BL, and NN agar plates for anaerobic sporeformers and TS agar for aerobic sporeformers. After incubation for 2 days, 15 bacterial groups, yeasts and molds were recognized by colonial and cellular morphologies, gram-reaction, spore formation, and aerobic and anaerobic growth.
[d]Bacterial number/g of wet feces.

Remarks

1. No significant changes were observed among the total numbers of fecal bacteria before, during, and after the feeding of *B. longum*.
2. The percentage of *Bifidobacterium* number to the total bacterial counts increased rapidly in the first week of the feeding experiment and maintained at a relatively high level throughout the feeding period. Its number decreased gradually after the subjects had stopped taking the *B. longum* preparation.
3. Proportion of *Bacteroides* and lecithinase-negative clostridia were markedly decreased by the feeding. Lecithinase-negative clostridia, e.g., *C. paraputrificum* and *C. innocuum* were found to constitute high proportion of fecal microflora in patients with colon cancer and implicated in the formation of carcinogens from bile acids.

Effect of *B. longum* Feedings on Fecal Properties

Duration of Feedings	Feces			
	Water Content (%)[a]	pH Value[b]	Ammonia Concentration (μg/dl)[c]	β-Glucuronidase Activity (μmol/h/g)[d]
Before feedings				
Second week	82.3 ± 3.1	6.0 ± 0.5	409.7 ± 124.0	11.4 ± 8.5
First week	81.9 ± 4.6	6.1 ± 0.3	422.1 ± 129.6	16.0 ± 6.3
During feedings				
First week	81.7 ± 5.2	6.1 ± 0.5	182.1 ± 98.5[e]	NT
Second week	79.6 ± 4.0	6.4 ± 0.5	242.9 ± 74.5[e]	NT
Third week	82.1 ± 4.7	5.9 ± 0.2	189.4 ± 69.7[f]	NT
Fourth week	79.2 ± 8.2	5.6 ± 0.4	172.6 ± 88.8[f]	9.2 ± 4.8
Fifth week	83.5 ± 4.3	5.6 ± 0.3[e]	174.6 ± 66.9[f]	6.1 ± 2.4[e]

(continued)

Effect of *B. longum* Feedings on Fecal Properties (*cont.*)

Duration of Feedings	Feces			
	Water Content (%)[a]	pH Value[b]	Ammonia Concentration (μg/dl)[c]	β-Glucuronidase Activity (μmol/h/g)[d]
After feedings				
First week	79.7 ± 4.4	6.1 ± 0.3	283.6 ± 45.5	14.5 ± 5.6
Second week	80.7 ± 3.4	6.1 ± 0.3	314.1 ± 78.8	13.8 ± 4.5

[a]Ten grams of feces were dried to constant weight to determine the water content.
[b]Fecal pH values were directly measured by a glass electrode.
[c]Ammonia concentration was determined by an ammonia test kit (Wako Pure Chemical Industries Ltd., Osaka, Japan).
[d]Fecal specimens were suspended in 25 mM phosphate buffer (pH 6.8), disrupted anaerobically by sonication at 4°C, and then centrifuged at 10,000*g* for 10 min. The supernatant was incubated anaerobically at 37°C for 30 min with a substrate solution of 10 mM phenolphthalein-β-D-glucuronide (Sigma Chemical Co., St. Louis, MO). After incubation, the reaction was stopped by adding 0.2 M glycine-NaOH buffer (pH 10.4) containing 0.2 M NaCl and the β-glucuronidase activity was estimated using photometrical absorbance at 540 nm.
[e]Statistically significant at $p < 0.05$ when compared with the values obtained before feedings.
[f]Statistically significant at $p < 0.01$ when compared with the values obtained before feedings.

Remarks Oral supplement of *B. longum* may be introduced to suppress the growth of lecithinase-negative clostridia and bacteroides, resulting in lower fecal pH, ammonia concentration, and β-glucuronidase activity.

5.5.3 Effect of *Bifidobacterium* in Patients with Gastrointestinal Disorder (7)

Effect of Feeding *Bifidobacterium* on Fecal Bacterial Flora[a]

	Before[b] (2 weeks)	During[c] (2 or 3 weeks)	After[d] (2 weeks)
Total anaerobes[d]	10.40 ± 0.10	10.45 ± 0.07	10.50 ± 0.16
Total *Bifidobacterium*[e]	9.63 ± 0.16	10.03 ± 0.07[g]	9.20 ± 0.31
B. breve strain Yakult	Not detected	7.99 ± 0.29	6.73 ± 0.31[f]
B. bifidum strain Yakult	Not detected	7.38 ± 0.26	5.32 ± 0.40[g]
Coliform[h]	8.10 ± 0.30	7.43 ± 0.42	7.76 ± 0.42
Total other aerobes[h]	7.69 ± 0.35	7.25 ± 0.38	7.04 ± 0.47[f]

[a]Data are mean ± SE of log number of bacteria per gram feces.
[b]There were 14 patients with digestive diseases: viz. one case each of postoperative intestinal adhesions, hemangioma of colon, acute enteritis, protein losing enteropathy complicated by malabsorption, diverticulosis of colon, short bowel syndrome complicated by liver cirrhosis, and liver cirrhosis complicated by ulcerative colitis; two cases each of chronic hepatitis and liver cirrhosis, and three cases of alcoholic liver disease. There were nine males and five females, ranging between 31 and 80 years of age (mean ± SEM, 50 ± 3). None received antibiotics. The patients were examined weekly.
[c]The BBG preparation was provided by Yakult Co. Ltd., Tokyo, containing viable *B. breve* strain Yakult

and *B. bifidum* strain Yakult at 1×10^9 each organism per gram. Each patient received orally 3 or 6 g of BBG daily for 2 or 3 weeks and was examined weekly.

[d]To 1 g of fresh feces, 9 ml of 0.1% corn steep liquor was added and mixed anaerobically. Supernate of it was further diluted with the same diluent under anaerobic condition. The total anaerobic bacterial counts were determined by cultivation on VL-Mix medium (Nissui Co. Ltd., Tokyo) for 48–72 h.

[e]Total number of bifidobacteria was determined by culturing on MPN medium containing streptomycin (3000 μg/ml) and neomycin (1000 μg/ml). The number of *B. bifidum* strain Yakult was counted on MPN medium supplemented with streptomycin (3000 μg/ml), neomycin (1000 μg/ml), and erythromycin (100 μg/ml). The difference between these two cultures was regarded as the number of *B. breve*.

[f]$p < 0.05$ as compared to the value before treatment.

[g]$p < 0.01$ as compared to the value before treatment.

[h]Aerobic cultures were set up on MacConkey's medium and incubated at 37°C for 24 h for enumeration of coliform bacteria and other aerobes.

Remarks Fecal bacterial flora showed no conspicuous alterations in total anaerobic bacterial counts during the BBG therapy. However, the total number of bifidobacteria in feces increased significantly and the coliform bacteria and other aerobes showed a tendency to decline. Both species of bifidobacteria were recovered in feces 2 weeks after completion of the treatment. Symptomatic discomforts including abdominal distension and pain, anorexia, edema, and fever of undetermined origin were reduced during the treatment.

5.5.4 Effect of *Bifidobacterium* and *L. acidophilus* in Patients with Leukemia (3)

Changes in Intestinal Bacterial Flora during Antileukemic Therapy

		Treatment with Antileukemic Drug[c]	
Intestinal Flora[a]	Control[b]	Without Bifidus[d]	With Bifidus[e]
Aerobic			
E. coli	7[f]	11[f]	8[f]
L. acidophilus	11	11	11
S. faecalis	7	6	8
Anaerobic			
Bacteroides	9	12	10
Bifidobacterium	7	7	11
Veillonella	9	11	10

[a]The intestinal bacteria of 56 patients with leukemia were isolated from the feces and identified by Mitsuoka's method (5) and Bergey's classification (2).

[b]Patients with leukemia who were not treated with antileukemic drugs.

[c]Drugs used in the patients for treatment of leukemia were chiefly: ordinary dosage of daunomycin, cytosin arabinoside, 6-MP, prednisolone, etc. Treatment was continued for more than 2 months in all cases.

[d]Patients treated with antileukemic drugs given plain drinking water.

[e]Patients treated with antileukemic drugs were given milk containing Morinaga Bifidus. One milliliter of milk contained about 10^7 of *Bifidobacterium* and *L. acidophilus*. Two hundred milliliters of milk was given to the patients everyday in this experiment and the effect was assessed after 3 months.

[f]Mean $\log_{10}$ number of intestinal bacteria per gram feces.

Remarks Change in intestinal flora of leukemia patients occurred during administration of antileukemic drugs with increase in the numbers of the various bacteria. However, with oral administration of *Bifidobacterium,* the intestinal bacterial balance was restored to normal. It is the upset in the balance of normal flora that makes leukemic patients prone to infectious diseases.

Changes in Minor Members of Intestinal Flora During Antileukemic Therapy

		Treatment with Antileukemic Drugs (56 cases)[b]	
Organism[a]	Control (10 cases)	Without Bifidus[c] (28 cases)	With Bifidus[c] (28 cases)
Klebsiella	0	3[d]	2[d]
Citrobacter	0	6	5
Proteus vulgaris, Pseudomonas, etc.	0	8	2

[a]Minor members of intestinal bacteria.
[b]Cases in which more than 10^6 colonies/g feces were observed.
[c]Same as previously mentioned for the preceding table.
[d]Mean $\log_{10}$ number of intestinal bacteria per gram feces.

Remarks In patients undergoing antileukemic therapy without *Bifidobacterium* administration, the numbers of the minor members of the intestinal flora is higher. However, *Bifidus* administration in the other group helped to decrease the number of minor members of the intestinal flora.

Changes in Yeast *Candida* in Feces During Antileukemic Therapy

No. of *Candida* per g Feces	Administration of Bifidus[a] (16 cases)		Nonadministration of Bifidus[b] (control, 11 cases)
	Before (cases)	After (cases)	
10^8	4	0	3
10^7	6	1	3
10^6	4	4	3
10^5	2	3	2
$<10^4$	—	8	0

[a]Oral administrations of milk containing Bifidus to 16 leukemic patients with more than 10^5 or more *Candida* per gram of feces.
[b]Eleven patients receiving antileukemic therapy without administration of milk containing Bifidus, and these patients also had 10^5 or more *Candida* per gram of feces.

Remarks *Candida* counts in feces of patients with leukemia were very high. Bifidus administration decreased the number of *Candida* in the leukemic patients, thus preventing *Candida* overgrowth in the intestines. The number of *Candida* remained the same in the control patients not administered with Bifidus.

Changes in Urine Indican and Blood Endotoxin During Antileukemic Therapy

	Urine Indican[a]		Blood Endotoxin[a]	
	No. of Cases Tested	No. of Positive Cases	No. of Cases Tested	No. of Positive Cases
Control	5	0	5	0
Patient				
Before treatment	7	1	5	1
Treatment with antileukemic drug	7	4	9	6
Antileukemic drug + *Bifidobacterium* administration	7	1	6	3

[a]Endotoxin in blood and indican in urine are metabolic substances of intestinal bacteria, which are often present in leukemic patients.

Remarks Both urine indican and blood endotoxin were inhibited by Bifidus administration. The levels of these substances were normalized as a result of the improvement of intestinal bacteria by Bifidus administration.

REFERENCES

1. Benno, Y., and Mitsuoka, T. Impact of *Bifidobacterium longum* on human fecal microflora. *Microbiol. Immun.* **36**(7): 683–694, 1992.
2. Buchanan, R. E., and Gibbons, N. E. (eds.). *Bergey's Manual of Determinative Bacteriology,* 8th ed. Williams & Wilkins, Baltimore, 1974.
3. Kageyama, T., Tomoda, T., and Nakano, Y. The effect of *Bifidobacterium* administration in patients with leukemia. *Bifidobact. Microflora* **3:** 29–33, 1984.
4. Lidbeck, A., Gustafssom, J.-A., and Nord, C. E. Impact of *Lactobacillus acidophilus* supplements on the human oropharyngeal and intestinal microflora. *Scand. J. Infect. Dis.* **19:** 531–537, 1987.
5. Mitsuoka, T. *A Color Atlas of Anaerobic Bacteria.* Sobunsha, Tokyo, 1980.
6. Spanhaak, S., Havenaar, R., and Schaafsma, G. The effect of consumption of milk fermented by *Lactobacillus casei* strain Shirota on the intestinal microflora and immune parameters in humans. *Eur. J. Clin. Nutr.* **52:** 1–9, 1998 (in press).
7. Tamura, N., Norimoto, M., Yoshida, K., Hirayama, C., and Nakai, R. Alteration of fecal bacterial flora following oral administration of bifidobacterial preparation. *Gastroenterol. Jpn.* **18**(1): 17–24, 1983.

5.6 BACTERIAL TRANSLOCATION

5.6.1 Introduction

Bacterial translocation is defined as the passage of viable bacteria from the gastrointestinal (GI) tract to extraintestinal sites, such as the mesenteric lymph nodes (MLNs), spleen, liver, kidneys, and peripheral blood. Bacterial translocation from the GI tract occurs when: (a) the mucosal epithelial barrier is damaged, (b) host defense mechanisms are suppressed, and (c) enteric flora are disrupted allowing overgrowth of certain bacterial species.

As to the roles of host defense systems in translocation, serum immunity does not appear to be effective in preventing the initial translocation of bacteria from the GI tract across the mucosal barrier to the MLNs but does effectively inhibit the systemic spread of translocating bacteria from the MLN to other organs and sites. T-cell-mediated immune responses appear to be important in immune defense against bacterial translocation. The incidence of spontaneous translocation is much higher in athymic nude (nu/nu) mice than in euthymic (nu/+) mice injected once intraperitoneally with imunosuppressive agents such as cyclophosphamide, prednisolone, methotrexate, 5-fluorouracil, or cytosine arabinoside, exhibiting increased bacterial translocation to the MLN, spleen, and liver. Vaccination with formalin-killed *Propionibacterium acnes* (formerly classified as *Corynebacterium parvum*), an immunomodulator, reduced both the incidence of translocation and the number of translocated *E. coli* in antibiotic-decontaminated mice infected with *E. coli* alone. Neither particulate yeast glucan nor muramyldipeptide inhibited the translocation in this murine model.

5.6.2 Effect on Lethal Bacterial Translocation in a Murine Model (1)

Prevention of Fluorouracil (5-FU)-induced Translocation by *L. casei* Strain Shirota

Group[a]	Survival Rate[b]	Log_{10} No. of *E. coli* in[c]		No. of WBC[d]
		Liver	Cecum	($10^3/mm^3$)
Untreated control	—	< 2 (0)	4.9 ± 0.2	44 ± 16
5-FU control	1/10	9.2 ± 0.6(100)	7.6 ± 0.5	4 ± 1
5-FU + LC 9018 (sc)[e]	18/20[f]	5.5 (20)	5.0 ± 0.8[f]	52 ± 37[f]
5-FU + LC 9018 (iv)[e]	17/20[f]	7.5 (10)	4.5 ± 0.7[f]	66 ± 34[f]

[a]Specific pathogen-free, 7-week-old male mice (BALB/c, C57BL/6, DBA/2, C3H/HeN, C3H/HeJ, etc.) were purchased from Charles River Japan Inc., Atsugi, Japan. The mice were maintained on MF diet (Oriental Yeast Co. Ltd., Tokyo) and sterilized water (126°C for 30 min), which contained Cl_2 at a concentration of 1.5 ppm, given *ad libitum.*

[b]Fluorouracil (Kyowa Hakko Kogyo Co. Ltd., Tokyo) was diluted in 0.7 M Tris-HCl buffer (pH 8.4) before use. 5-FU at a dose of 400 mg/kg (500 mg/kg for tumor-bearing mice) was administered orally (10–12 mice/group) and the mice were observed for survival for 40 days (55 days for tumor-bearing mice) after treatment.

[c]To determine the numbers of viable bacteria in organs, on day 12 after treatment with 5-FU, mice were killed by cervical dislocation under ether anesthesia and transferred to an anaerobic glove box. The peritoneal washing fluid was obtained by lavaging the peritoneal cavity with 5 ml of sterile buffer

(KH_2PO_4, 0.45%; Na_2HPO_4, 0.6%; L-cysteine hydrochloride, 0.05%; Tween-80, 0.05%; resazurin, 0.0001%; and agar, 0.1%), which had been reduced in the glove box before use. The liver, spleen, mesenteric lymph nodes, small intestine, cecum, and colon and rectum were removed, placed in grinding tubes containing 5 ml of the same buffer and homogenized with a Teflon grinder. The homogenates were diluted serially with the buffer, and the numbers of viable bacteria in the suspensions were determined using culture media. Values in parentheses are the percent of mice in which bacteria were detected.

[d]Peripheral blood was obtained from the mice on day 11 after administration of 5-FU. The concentrations of white blood cells (WBC) were counted.

[e]*L. casei* strain Shirota grown in Rogosa's medium was harvested and washed thoroughly with distilled water. The bacteria suspended in distilled water were heat killed at 100°C for 30 min and lyophilized. The lyophilized powder was suspended in saline and injected intravenously or subcutaneously at a dose of 80 mg/kg (sc) or 20 mg/kg (iv) on day 15, and in the experiment designed to observe survival on day 10.

[f]Significantly different from 5-FU-treated controls $p < 0.01$.

Remarks A single administration of fluorouracil (5-FU), a widely used chemotherapeutic agent for cancer, at high doses (338–800 mg/kg) to specific pathogen-free mice induced a lethal infection with *E. coli*. The infection manifested in all mice treated with 5-FU 7–14 days after administration of the drug, when the number of hepatic *E. coli* reached levels ranging from 10^8 to 10^{10} colony-forming units. The infection was accompanied by an increase in the population of *E. coli* in the intestinal tract, which reached about 10^3–10^4 the level of normal mice (3). Vaccination of mice with heat-killed *L. casei* strain Shirota (LC 9018) via an intravenous or subcutaneous route reduced the lethal toxicity of 5-FU at large doses and the occurrence of systemic infection with indigenous *E. coli*. Overgrowth of *E. coli* induced by large doses of 5-FU was also inhibited by pretreatment of mice with *L. casei* strain Shirota (Refs. 1, 2).

REFERENCES

1. Nomoto, K., Yokokura, T., Mitsuyama, M., Yoshikai, Y., and Nomoto, K. Prevention of indigenous infection of mice with *Escherichia coli* by nonspecific immunostimulation. *Antimicrob. Agents Chemother.* **36:** 361–367, 1992.
2. Nomoto, K., Yokokura, T., and Nomoto, K. Prevention of 5-fluorouracil-induced infection with indigenous *Escherichia coli* in tumor-bearing mice by nonspecific immunostimulation. *Can. J. Microbiol.* **38:** 774–778, 1992.
3. Nomoto, K., Yokokura, T., Yoshikai, Y., Mitsuyama, M., and Nomoto, K. Induction of lethal infection with indigenous *Escherichia coli* in mice by fluorouracil. *Can. J. Microhiol.* **37:** 244–247, 1990.

6

ENHANCEMENT OF INDIGENOUS PROBIOTIC ORGANISMS

PREBIOTICS

Prebiotics are nondigestible food ingredients that beneficially affect the host by selectively stimulating the growth and/or activity of one or a limited number of bacterial species already resident in the colon, and thus attempt to improve host health (7).

Lactic acid bacteria such as lactobacilli and bifidobacteria are among the predominant microorganisms in the intestine of humans and animals. It has been shown that the number of these indigenous lactic acid bacteria could decrease due to some physiological conditions, diseases, and aging, resulting in various gastrointestinal ailments. Prebiotics in food would pass through the stomach unaltered to reach the small intestine and eventually the large intestine, where the prebiotics could be selectively utilized by bifidobacteria and some lactobacilli. This could lead to an increase in the number of indigenous probiotic bacteria in the intestine.

TABLE 6.1 Slowly Absorbable Substrates and Their Effect on the Intestinal Microflora[a]

Carbohydrate Substrate	Benefactor	Sources	Daily Dose (g/kg body wt)	Safety Status	Major Manufacturers	Trade Names	Ref.
Oligosaccharides							
Fructooligosaccharides (1-kestose, nystose, fructofuranosylnystose)	Bifidobacteria, bacteroides	Jerusalem artichoke tubers β-fructofuranosidase conversion from sucrose/inulin chemical synthesis	0.3	Listed in FOSHU 1991	Meiji Seika Kaisha (Japan) Beghin-Meiji Industries (France) Golden Technologies (USA) Cheil Foods and Chemicals (Korea) ORAFTI (Belgium) Cosucra (Belgium)	Meioligo Actilight NutraFlora Oligo-Sugar Raftilose and Raftiline Fibruline	2, 6, 9
4′ Galacto-oligosaccharides	Bifidobacteria, lactobacilli	Human/cow milk β-D-galactosidase conversion from lactose	0.128	Listed in FOSHU 1991	Yakult Honsha (Japan) Nissin Sugar Manufacturing Co. (Japan) Snow Brand Milk Products (Japan) Borculo Whey Products (the Netherlands)	Oligomate Cup-Oligo P7L and others TOS-Syrup	2, 10
4′ Galactosyl-lactose	Bifidobacteria	Human milk	0.3				14
Galactosyl-oligosaccharides	Bifidobacteria		0.3				13
Palatinose (6-0-α-D-glucopyranosyl-D-fructoruranose)	Bifidobacteria	Palatinose synthase conversion of sucrose Intermolecular condensation of glucose		Listed in FOSHU 1991	Mitsui Sugar Co. (Japan)	ICP/O IOS	2, 11

Raffinose	Bifidobacteria	White beet	0.45				1
Soybean oligosaccharides (stachyose/raffinose-sucrose)	Bifidobacteria, some lactobacillus	Soybean/whey,		Listed in FOSHU 1991	The Calpis Food Industry Co. (Japan)	Soya-oligo	2, 8, 17
Transgalactosyl oligosaccharides (TOS)	Bifidobacteria						12, 16
Xylo-oligosaccharides	Bifidobacteria	Controlled enzymatic hydrolysis by endo 1,4-β-xylanase	0.12	Listed in FOSHU 1996	Suntory Ltd. (Japan)	Xylo-oligo	2, 15
Disaccharides and polyols							
Lactulose (4-0-α-D-galactopyranosyl-D-fructofuranose	Bifidobacteria Lactic acid bacteria	Catalytic hydrogenation of lactose at high temperature	0.15	Listed in FOSHU 1996	Morinaga Milk Industry Co. (Japan) Solvay (Germany) Milei GmbH (Germany) Canlac Corporation (Canada) Laevosun (Austria) Inalco SPA (Italy)	MLS/P/C	2, 18, 20
Lactitol (1,4-galactosylglucitol)		Catalytic hydrogenation of lactose/lactulose at high temperature		ADI not specified (WHO, 1982)			4, 21
Lactosucrose	Bifidobacteria	Transfrutosylation of lactose and sucrose by β-fructofuranosidase	—	Listed in FOSHU 1996	Ensuiko Sugar Refining Co. (Japan) Hayashibara Shooji Inc. (Japan)	Nyuka-Origo Newka-Oligo	2, 5

(continued)

TABLE 6.1 Slowly Absorbable Substrates and Their Effect on the Intestinal Microflora[a] (*cont.*)

Carbohydrate Substrate	Benefactor	Sources	Daily Dose (g/kg body wt)	Safety Status	Major Manufacturers	Trade Names	Ref.
Disaccharides and polyols (*cont.*)							
Mannitol	Gram-negative flora	Ash, figs, olives, algae Catalytic hydrogenation of mannose/fructose	0.17	ADI not specified (WHO, 1982) GRAS (FASEB/SCOGS, 1973)			3, 19, 21
Sorbitol	Gram-negative flora	Cherries, pears, cider Catalytic hydrogenation of glucose inverted sugar	0.17	ADI not specified (WHO, 1982) GRAS (FASEB/SCOGS, 1973)			3, 19, 21
Xylitol	Gram-positive acid-producing flora	Raspberries, mushroom, algae Catalytic hydrogenation of xylase	4.5	ADI not specified (WHO, 1982)			19, 21

[a]FOSHU = Food for Specified Health Use legislated in Japan; WHO = World Health Organization; GRAS = Generally recognized as safe (United States); ADI = Acceptable daily intake.

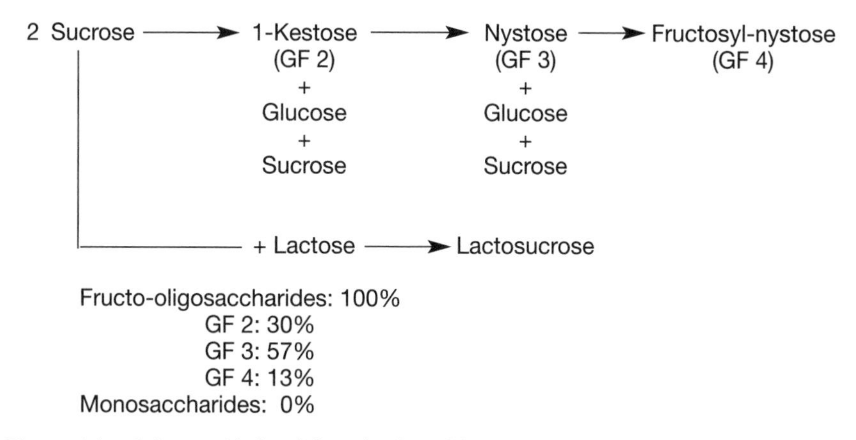

Figure 6.1. Scheme of industrial production of fructo-oligosaccharides and lactosucrose by an enzymatic process. The enzyme β-fructo-furanosidase is obtained from *Aspergillus niger* (7).

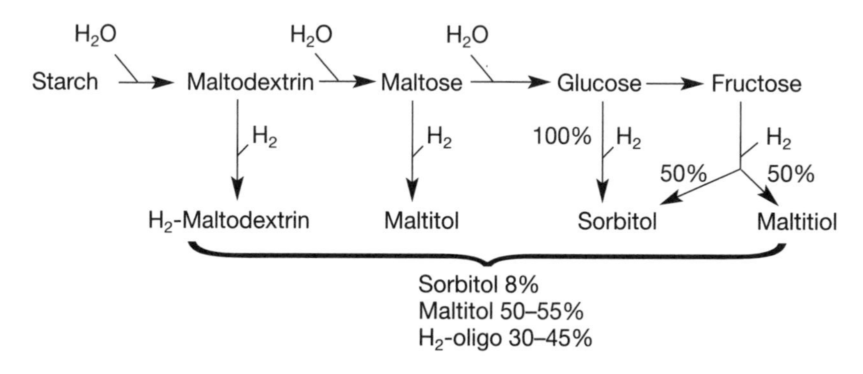

Figure 6.2. Scheme of industrial production of sugar alcohol from starch.

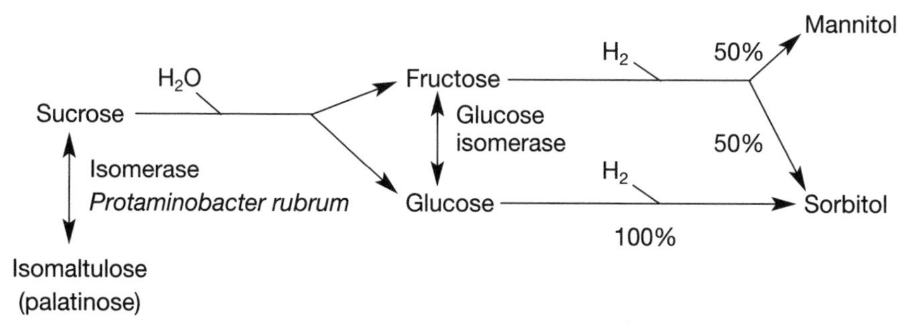

Figure 6.3. Scheme of industrial production of sugar alcohol from sucrose.

REFERENCES

1. Benno, Y., Endo, K., Shiragmai, N., Sayanama, K., and Mitsuoka, T. Effect of raffinose on human fecal microflora. *Bifidobacteria Microflora.* **6:** 59–63, 1987.
2. Crittendon, R. G., Playne, M. J. Production, properties and application of food-grade oligosaccharides. *Trends Food Sci. Technol.* **7:** 353–361, 1996.
3. FASEB/ SCOGS. Evaluation of sorbitol and mannitol as food ingredients. *NTIS PB. Rep.* 9–10, 221–951, 953, 1973.
4. Felix, Y., Hudson, M., Owen, R., Ratcliffe, B., van Es A., van Velthuijsen, J., and Hill, M. Effect of dietary lactitol on the consumption and metabolic activity of the intestinal microflora in the pig and human. *Microb. Ecol. Health Dis.* **3:** 259–267, 1990.
5. Fujita, K., Kitahata, S., Kozo, H., and Hotoshi, H. Production of lactosucrose and its properties. In *Carbohydrates in Industrial Synthesis (Proceedings of the Symposium of the Division of Carbohydrate Chemistry of the American Chemical Society).* Ed. Clarke, M. A. pp. 68–76, Bartens, Berlin, 1992.
6. Gibson, G. R., Beatty, E. R., Wang, X., and Cummings, J. H. Selective stimulation of Bifidobacteria in the human colon by oligofructose and inulin. *Gastroenterology* **108:** 975–982, 1995.
7. Gibson, G. R., and Roberfroid, M. B. Dietary modulation of the human colonic microbiota: Introducing the concept of prebiotics. *J. Nutr.* **125**: 1401–1412, 1995.
8. Hayakawa, K., Mizutani, J., Wada, K., Masai, T., Yoshihara, I., and Mitsuoka, T. Effects of soybean oligosaccharides on the human fecal flora. *Microb. Ecol. Health Dis.* **3:** 293–303, 1990.
9. Hidaka, H., Eida, T, Takizawa, T., Tokunaga, T., and Tashiro, Y. Effects of fructooligosaccharides on intestinal flora and human health. *Bifidobact. Microflora* **5**(1): 37–50, 1986.
10. Ito, M, Deguchi, Y., Miyamori, A., Matsumoto, K., Kikuchi, H., Matsumoto, K. Kobayashi, Y., Yajima, T., and Kan, T. Effects of administration of galactooligosaccharides on the human fecal microflora, stool weight and abdominal sensation. *Microb. Ecol. Health Dis.* **3:** 285–292, 1990.
11. Kashimura, J., Nakajima, Y., Benno, Y., Endo, K., and Mitsuoka, T. Effects of palatinose and its condensate intake on human fecal microflora. *Bifidobact. Microflora* **8:** 45–50, 1989
12. Kohmoto, T., Fukui, F., Takaku, H., Machita, Y, Arai, M., and Mitsuoka, T. Effect of isomaltito-oligosaccharides on human fecal flora. *Bifidobact. Microflora* **7:** 61–69, 1988.
13. Minami, Y., Kouchei, Y., Tamura, Z., Tanaka, T., and Yamamoto, T. Selectivity of utilization of galactosyl-oligosaccharides by bifidobacteria. *Chem. Pharmac. Bull.* **31:** 1688–1691, 1983.
14. Ohtsuka, K., Benno, Y., Endo, K., Uedo, H., Ogawa, O., Uchida, T., and Mitsuoka, T. Effect of 4-galactosyllactose intake on human fecal microflora. *Bifidus* **2:** 143–149, 1989.
15. Okazaki, M., Fujikawa, S., and Matsumoto, N. Effect of xylooligosaccharide on the growth of bifidobacteria. *Bifidobact. Microflora* **9:** 77–86, 1990.
16. Rowland, I. R., and Tanaka, R. The effects of transgalactosylated oligosaccharides on gut flora metabolism in rats associated with a human faecal microflora. *J. Appl. Bacteriol.* **74:** 667–674, 1993.
17. Saito, Y., Iwanami, T., and Rowland, I. The effects of soybean oligosaccharides on the

human gut microflora cultured in *in vitro* culture. *Microb. Ecol. Health Dis.* **5:** 105–110, 1992.

18. Salminen, S., and Salminen, E. Lactulose and lactitol induced caecal enlargement and microflora changes in mice. In *European Food Toxicology. II.* Ed. Battaglia, R., pp. 313–317. Institute of Toxicology, ETH, Zurich, 1986.
19. Salminen, S., Salminen, E., Bridges, J., and Marks, V. Gut microflora interactions with xylitol in the rat, mouse and man. *Food Chem. Toxicol.* **23:** 985–990, 1985.
20. Terada, A., Hara, H., Kataoka, M., and Mitsuoka, T. Effect of lactulose on the composition and metabolic activity of the human fecal flora. *Microb. Ecol. Health Dis.* **5:** 43–50, 1992.
21. World Health Organization (WHO). *Toxicological Evaluation of Certain Food Additives,* WHO Food Addit. Ser. W.H.O., Washington, DC, 1982.

INDEX